BASICS OF
AUDIO AND VISUAL SYSTEMS DESIGN

Raymond H. Wadsworth, P.E., is Senior Vice-President for Hubert Wilke, Inc., communications facilities consultants. For nine years, prior to joining the Wilke organization in 1965, Mr. Wadsworth was Director of Engineering and Product Development for TelePrompTer Corporation and TelePro Industries. During that time he supervised design and engineering for over 100 audio-visual system installations, including the Air Force Command Post, Joint Chiefs of Staff in the Pentagon, National Aeronautic and Space Administration, White Sands Missile Range, Johnson & Johnson, Consolidated Edison, Phillips Petroleum Company, Princeton University, University of Illinois, and Orange Coast College.

He received his Mechanical Engineering Degree from Duke University and his Professional Engineer's License from the University of the State of New York.

He is co-author of an architectural textbook and has written extensively for such magazines as ''Machine Design,'' ''Architectural Record,'' ''American School and University,'' and ''Video Systems.''

The National Audio-Visual Association (NAVA) takes pride in serving the industry that is leading the communications revolution of America. NAVA is the international trade organization of the audio-visual communications industry, including video, microcomputer, and other advanced technology. NAVA members are business and media professionals concerned with developing and promoting more advanced and effective ways of educating, communicating, and training.

In addition, hundreds of individuals hold associate membership in NAVA; these persons use A-V products for training, education, and information dissemination.

****

The NAVA Industry and Business Council (I&BC) is one of seven NAVA councils formed to represent the various segments of the audio-visual communications industry. The I&BC coordinates activities of dealers, manufacturers, and users of audio-visuals in the industry and business market. The I&BC stimulates the development of new communications hardware, software, and services through projects such as the semi-annual Institute for Effective Communications during which business executives learn better ways of solving company problems. The I&BC, under the leadership of Chairman Robert P. Abrams, CMS (Williams, Brown, & Earle, Philadelphia, PA), sponsored the project which resulted in the publication of *Basics of Audio and Visual Systems Design.*

BASICS OF
AUDIO AND VISUAL SYSTEMS DESIGN

by

Raymond H. Wadsworth, P.E.

Howard W. Sams & Co., Inc.
4300 WEST 62ND ST. INDIANAPOLIS, INDIANA 46268 USA

166200

FIRST EDITION
FIRST PRINTING—1983

International Standard Book Number: 0-672-22038-5
Library of Congress Catalog Card Number: 82-61968

Edited by: *Jim Rounds*

Printed in the United States of America.

Preface

This is a basic text, dealing with fundamental concepts and procedures in the design of audio-visual systems. It is written primarily for those who are connected with the audio-visual industry in any of the many categories that fall within its scope.

The book contains information vital to the system designer, the architect, the contractor, the equipment supplier, the user, the system draftsperson, the student, the teacher, and the consultant.

There are newcomers to the field who have not had the advantage of several years of experience and find themselves working with out-dated information and rules of thumb that have somehow lost the reason for their origin. Still others, although actually connected with the industry for many years, have not been keeping up with the present-day technology.

It is hoped that readers in all of the preceding categories will find valuable information within these pages. The author has tried to maintain technical accuracy and integrity, without resorting to other than fundamental mathematics, and explanations are as concise as is consistent with clarity. A question and answer format was chosen specifically to accelerate recognition of important concepts. Examples are given when necessary to illustrate certain procedures, and explanatory figures are used throughout.

Information in tabular form has been presented to lessen the need for mathematical calculations, but where formulas are appropriate, the author has been careful to give the required units of measurement and show practical applications. The book is organized in sections, and carries a complete index for rapid access to a desired subject.

It would require a book many times the size of this to do justice to all the facets of the subject, hence superficial coverage of a wide range of subjects has given way to a more thorough treatment of those fundamentals which form the foundation for continued growth in this exciting field.

In preparing this book, the author draws from a wide experience he gained in over three decades of participation in audio-visual technology—in theory, equipment design, system design, sales, installation, teaching, and for the past 16 years, consulting. It is sincerely hoped that this text will prove helpful to its readers, and fulfill the purpose for which it was written.

RAYMOND H. WADSWORTH, P.E.

NAVA and the Industry and Business Council express their appreciation to the following individuals whose enthusiasm, ideas, work, and negotiations made possible the publication of this book: Robert P. Abrams, CMS, (Williams, Brown, & Earle, Philadelphia, PA), Chairman of the NAVA Industry and Business Council; Harry R. McGee, CAE, NAVA Executive Vice President; Kenton Pattie, NAVA Senior Staff Vice President; Kathleen M. Springer, NAVA Publications and Marketing Consultant.

Contents

INTRODUCTION

It has been said that our sense of sight accounts for 70% of what we learn; our sense of hearing, 20%; and our other senses of smell, taste, touch, and sensation, account for the remainder. In view of this we can agree that the visual display deserves the lion's share of our design efforts. When talking about visual displays, we are of course, including a wide spectrum of the communications media — print, photo, drawings, artwork, models, computer display, the projected image, and the televised image. However, it is with the last two kinds of display that we are concerned here. We will, therefore, start our discussion with projection systems.

The systems we will cover are those that are, in general, audience oriented. In other words, we will exclude all of the sales and promotional and one-on-one portable units that make use of talking slides, films, cartridges, and audio cassettes. These are specialties that require no design effort on our part, and can be used in an office, a classroom, a library, a study carrel, or virtually anywhere there is a viewer and a source of power.

We are going to discuss the design of *projection systems* for classrooms, conference rooms, board rooms, training centers, meeting rooms, small auditoriums, and the like. A *projection system* recognizes the importance of the interdependence between the basic elements that comprise the system. These elements are:

- the projector
- the projection screen
- the sound system
- the control system

In discovering their interdependence, we will see that a projection system does not consist merely of a group of unrelated components. But rather, the combining of these components into an integrated system, in such a way that each is compatible with the other.

There are two additional elements which are not in the category of components. The first, and most important, is the viewing audience, and the second is the environment. A good portion of our design work must consider the audience. Viewers must see, they must hear, and they must be comfortable. Seeing means good sight lines, proper image sizes, sufficient screen brightness, and compatible image graphics. Hearing means high-quality sound, with good speech intelligibility and music fidelity, with proper volume control and directivity. Comfort means proper lighting for no-glare see-ability, ideal air temperature, humidity, and air distribution with low noise level, comfortable seating, and pleasing decor. Successful facilities are those wherein careful attention has been paid to the integrated design of all elements — in short, the systems approach.

Fig. 1-1

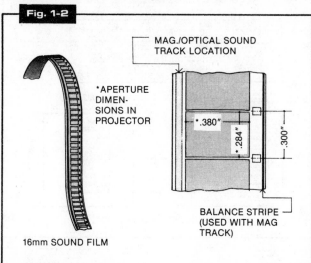

Fig. 1-2

MAG./OPTICAL SOUND TRACK LOCATION

*APERTURE DIMEN-SIONS IN PROJECTOR

*.380"

*.284"

.300"

BALANCE STRIPE (USED WITH MAG TRACK)

16mm SOUND FILM

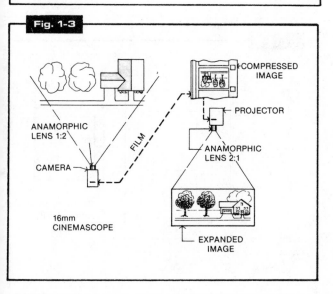

Fig. 1-3

COMPRESSED IMAGE

PROJECTOR

ANAMORPHIC LENS 1:2

FILM

CAMERA

ANAMORPHIC LENS 2:1

16mm CINEMASCOPE

EXPANDED IMAGE

THE PROJECTED IMAGE FORMAT

Probably because man's vision is oriented horizontally, and he usually sees more in the length of his horizon as depicted in Fig. 1-1 than he does in the height of the objects on it, it is a natural tendency to want to frame a scene horizontally. Although the photographic process was well established, professional film widths and frame sizes were not standardized until the early 1900s and the 35-mm width was accepted as standard in Europe and in the United States. Images were thus framed in the approximate 4 to 3 ratio of width to height, and this shape was standardized by the Academy of Motion Picture Arts and Sciences in 1927, and became known as the "Academy" format. Each frame measured 0.825 in wide by 0.600 in high, and spanned 4 sprocket holes on the film.

Since that time there have been many variations in the frame size of various formats. Film widths as wide as 70-mm became popular in 1952, accompanied by a dozen or more formats to accommodate stereophonic sound, with optical and magnetic sound tracks. Today, the original 35-mm Academy format is used only for newsreels, certain films for television transmission, industrial films, and documentary films.

Unlike the professional theater film formats mentioned above, the half-size 16-mm film format, shown in Fig. 1-2, which became popular in the 1920s, has remained unchanged except for the addition of the sound track. More recent development of anamorphic optics has made 16-mm cinemascope pictures practical, using the basic "Academy" format of 0.380 in wide by 0.284 in high, in conjunction with a 2:1 expander lens (or anamorphic) attachment to the regular prime lens.

In cinemascope photography, anamorphic optics are used on the camera to *compress* the image horizontally by a factor of two. The compression takes place in the horizontal direction only, squeezing twice the information into the single frame width of 0.380 in on the 16-mm film. When projected with an anamorphic attachment added to the projection lens, the image is *expanded* by a factor of two, producing the widescreen cinemascope image shown in Fig. 1-3.

WHAT IS THE ASPECT RATIO OF THE STANDARD ACADEMY FORMAT 16-MM FILM IMAGE?

The *aspect ratio* is defined as the width of the image divided by its height. For the 16-mm film image, this is

$$A.R. = \frac{0.380''}{0.284''} = 1.33:1, \text{ or 4 to 3}$$

WHAT IS THE ASPECT RATIO OF THE 16-MM PROJECTED CINEMASCOPE IMAGE?

$$A.R. = \frac{0.380'' \times 2}{.284} = 2.66:1$$

We note that the term aspect ratio can be applied to any rectangle, whether it be the actual film image, the projected screen image, or the actual aperture openings in the camera and projector.

WHAT IS THE SIZE AND ARRANGEMENT OF THE IMAGE ON A FILM STRIP?

The final positive film strip image is printed from a negative film, wherein the frame size is the camera aperture size. The camera aperture size is always slightly larger than the projector aperture, to ensure that there is no loss of information around the edges of the frame. In the case of the film strip, the camera records the image with a frame size of 0.965 in × 0.709 in, and the final transparent positive print is printed with a black masked frame measuring 0.885 in wide × 0.668 in high, with a resulting aspect ratio of 1.33:1 (Fig. 1-4). This is the format projected onto the screen.

It should be noted that the film strip moves through the camera and the projector vertically, so that the long dimension of the image is across the width of the film. This is in contrast to the 35-mm slide, which for horizontally oriented images, is positioned horizontally in both the camera and the slide projector.

WHAT IS THE STANDARD 35-MM SLIDE FORMAT?

In order to provide a larger image area than available with a single frame format in the vertical direction of 35-mm film, cameras were designed to transport the film horizontally, thus taking advantage of the film width as the image height. To obtain a suitable long dimension, it was necessary to span two single frame spaces on the film, as show in Fig. 1-5, occupying eight sprocket holes on each side of the film. Hence, the slide format is correctly called "35-mm double frame," abbreviated 35-mm d.f.

The camera aperture produces frames measuring 1.429 in × 0.965 in, which, as previously explained, are slightly larger than the frame on the slide mounts. The standard slide mount has a masked opening 1.34 in × 0.902 in, and may be used either horizontally or vertically in the projector, as the slide mount itself is perfectly square.

CALCULATE THE ASPECT RATIO OF THE 35-MM d.f. SLIDE

$$\text{A.R.} = \frac{1.34''}{0.902''} = 1.48{:}1$$

WHAT IS A SUPERSLIDE?

The superslide shown in Fig. 1-6 is a square format slide transparency that mounts in the standard 2- × 2-in mount. The masked aperture is 1½ × 1½ in. Although it has very attractive advantages, it has not found wide acceptance, for the simple reason that it cannot be produced by a 35-mm camera. A type of amateur camera that was ideal for making superslides was the 44- × 44-mm miniature dual-lens reflex, which used size 127 roll film. Unfortunately, most cameras that were designed for the 127 film size disappeared from the market when the 35-mm single-lens reflex camera became popular.

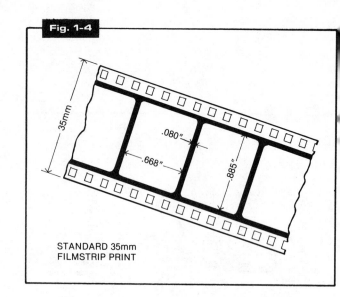

Fig. 1-4

STANDARD 35mm FILMSTRIP PRINT

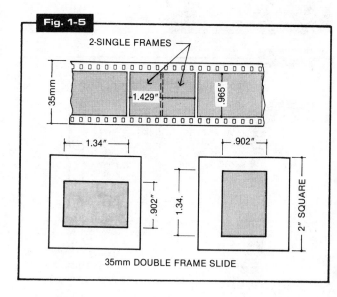

Fig. 1-5

2-SINGLE FRAMES

35mm DOUBLE FRAME SLIDE

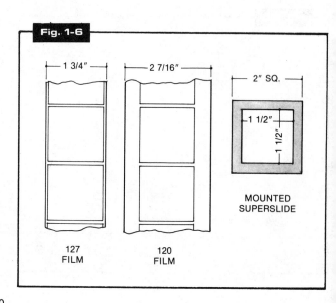

Fig. 1-6

127 FILM

120 FILM

MOUNTED SUPERSLIDE

Professionally, superslides are made on cameras using the well-known size 120 roll film, employing a superslide mask in the film case. The well-known (and expensive) Hasselblad camera is such a camera. Note that considerable film is wasted in trimming the 120 film to size. The aspect ratio is 1:1.

WHAT OTHER SLIDE MOUNTS ARE AVAILABLE?

"INSTAMATIC"®126 Format

This was the first cartridge type format to become popular in the miniature category suitable for both slides and prints. As shown in Fig. 1-7, slides are mounted in the standard 2- × 2-in mounts, masked for an opening of 1.04 × 1.04 in. The aspect ratio is 1:1, as are all square formats. "Instamatic" type cameras are necessary for its use.

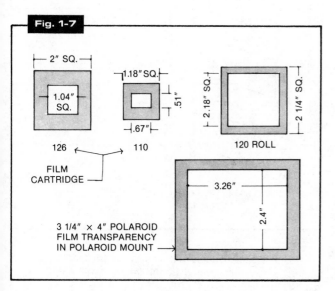

110 Format

With the introduction of some 18 new 110 format cameras during the last three years, along with the availability of several film types, and three miniature slide projectors, the 110 system has gained tremendous popularity. The film is composed of specially perforated 16-mm filmstock, packaged in convenient "drop-in" plastic cartridges.

Slide transparencies, Fig. 1-7, may be mounted in the small 1.18-in square (30-mm × 30-mm) plastic mount, made for the miniature slide projector, or in the standard 2- × 2-in mount with proper 0.51 × .67 in (13 mm × 17 mm) masked opening.

The 110 format has an aspect ratio of 1.31, not too different from the 1.48 ratio of the 35-mm slide. Consequently, it is possible to mount four 110 transparencies in a specially masked 2- × 2-in mount, to cover the single frame masked opening of the 35-mm d.f. slide. Such an arrangement is handy for certain multi-image techniques.

2¼- × 2¼-In Format

This format shown in Fig. 1-7 is produced on 120 roll film, used in a Rollei type twin lens camera, or with a Hasselblad type camera. This size slide is not popular and can be accommodated by only a few slide projectors. The carousel-type projectors will not accept this mount.

3¼- × 4-In Format

Special large format projectors are needed to handle this size transparency. The optics of these projectors will cover a masked opening up to 3 × 3¾ in, consequently the aspect ratio depends on the mask size. The Polaroid mount, designed to take the Polaroid positive transparency film, has an opening measuring 3.26 × 2.4 in, as shown in Fig. 1-7 with an aspect ratio of 1.36:1.

Overhead Projector Format

Inasmuch as the overhead (O.H.) projector projects transparencies, it may be thought of as a large slide projector. The mask sizes are not standardized, and the operator may use any size mask he or she desires, up to the maximum capacity of the projector. Most O.H. projectors are designed to accommodate masked copy up to 10 × 10 in, although the most popular mask size is probably 7½ × 9½ in. This represents an aspect ratio of 1.3:1, which is reasonably compatible with both the 35-mm slide and the 16-mm film formats.

Opaque Projector Format

There is no standardized format for the opaque projector. It projects by reflection of light from whatever opaque object is placed on its illuminated stage. Objects up to 10 × 10 in can be covered by the typical optical system. Due to the fact that the reflected light system can never be as bright as transmitted light, the opaque projector image appears its best when projected on a screen not wider than 4 or 5 ft. It is not a large audience device.

Television Format

The target area on television camera pick-up tubes is masked to an aspect ratio of 1.33:1, or 4:3. Consequently the televised image exhibits the same ratio. This means that television monitors and receivers, as well as large image tv projectors, all produce a 4 to 3 format image. This fact must be kept in mind when producing art work for tv transmission, or when planning slides to be televised. Standards must be followed which define the "safe area" for any material to be picked up by a television camera. Working within such limits will ensure that all the televised copy will appear on the tv receiver, or on the projection screen.

Table 1-1 recaps the various formats, the height and width of their projected apertures, their aspect ratios, and the corresponding relationship between the height and width of the projected image on the screen.

In this table, and throughout this text, lower case h and w will be used to represent the actual height and width of the projected aperture, whether or not it is formed by a photographed mask on the film (as in the case of the film strip), or by the physical opening in a slide mount, or by the aperture plate in the projector (as in the case of a 16-mm projector).

A capital H and W will indicate the screen image size, and will likewise be used throughout the text.

To make Table 1-1 complete, some additional formats are listed beyond those discussed above, namely 35-mm and 70-mm motion-picture film formats for professional use, and 8-mm and super 8-mm film formats. The 35-mm film formats are useful in large auditorium work, and the 8-mm formats in certain small room applications and laboratory uses. The 70-mm format is usually found only in large theater use.

® "Instamatic" is a registered trademark of the Eastman Kodak Company, Rochester, N.Y.

Table 1-1. Film and Slide Formats, Aspect Ratio, Aperture Size (In), Image Size (Ft)

FORMAT	Aspect Ratio	Aperture (Inches)		Image Size (Feet)	
		h	w	H	W
Column No. →	1	2	3	4	5
16-MM MOTION PICTURE	1.33	0.284	0.380	V ÷ 8	1.33H
16-MM CINEMASCOPE	2.66	0.284	0.380	V ÷ 8	2.66H
35-MM MOTION PICTURE	1.37	0.600	0.825	V ÷ 8	1.37H
35-MM WIDE SCREEN	1.85	0.446	0.825	V ÷ 6 TO V ÷ 8	1.85H
35-MM CINEMASCOPE	2.35	0.715	0.839	V ÷ 6 TO V ÷ 8	2.35H
70-MM WIDE SCREEN	2.21	0.868	1.913	V ÷ 6 TO V ÷ 8	2.21H
8-MM MOTION PICTURE	1.33	0.129	0.172	V ÷ 8	1.33H
SUPER 8-MM M.P.	1.34	0.158	0.211	V ÷ 8	1.34H
35-MM FILM STRIP	1.32	0.668	0.885	V ÷ 8	1.32H
35-MM d.f. SLIDE (HOR)	1.48	0.902	1.34	V ÷ 8	1.48H
35-MM d.f. SLIDE (VERT)	0.68	1.34	0.902	V ÷ 5.4	0.68H
SUPER SLIDE	1.00	1.50	1.50	V ÷ 5.4	H
"INSTAMATIC" 126	1.00	1.04	1.04	V ÷ 5.4	H
110 CARTRIDGE	1.31	0.51	0.67	V ÷ 8	1.31H
3¼" × 4" "LANTERN" SLIDE	1.09	2.75	3.00	V ÷ 8	1.09H
3¼" × 4" POLAROID SLIDE	1.36	2.40	3.26	V ÷ 8	1.36H
OVERHEAD TRANSPARENCY	VARIES	10" MAX	10" MAX	V ÷ 5.4	VARIES
OPAQUE PROJECTOR	VARIES	10" MAX	10" MAX	V ÷ 5.4	VARIES

h = Height of projected aperture, inches.
w = Width of projected aperture, inches.
V = Viewing distance, from rear corner viewer to diagonally opposite side of screen, feet.
H = Image height, feet.
W = Image width, feet.

WHAT DETERMINES THE PROPER SIZE PROJECTION IMAGES?

A projected image is sized correctly when the most distant viewer can read characters, numerals, and other symbols that have been prepared in accordance with proper graphic standards. Experience had taught us that graphics prepared in accordance with recommended standards could be read easily from a distance *six* times the *width* of the screen image. This determination, however, was made many years ago, when the average image had the common 4 to 3 aspect ratio. As images of varying widths came into use, it became apparent that the "6W rule" led to ambiguity, because it gave a different viewing distance for every different format width. Refer to Fig. 1-8 for example. Let us assume we can show an audience any one of four different image formats, all 5 ft 0 in high. These will be:

16-mm motion picture, W = 1.33 × 5' = 6.65'
35-mm slide, W = 1.48 × 5' = 7.4'
35-mm widescreen, W = 1.85 × 5' = 9.25'
16-mm cinemascope W = 2.66 × 5' = 13.3'

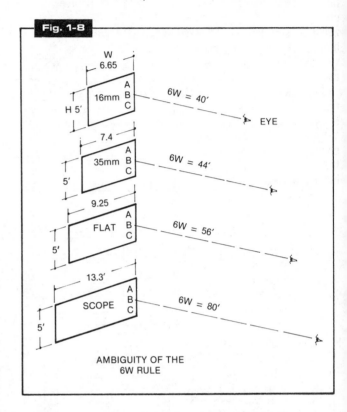

Fig. 1-8

AMBIGUITY OF THE 6W RULE

Let us further assume that there are 14 lines of caption material, here represented by the letters A, B, and C. (Fig. 1-8 is too small to show 14 lines of type, but the principle is still valid.)

Now according to the 6W rule, a viewer should not sit further from the screen than:

6 × 6.65' = 40.0' to view the 16-mm m.p.,
6 × 7.4' = 44.4' to view the 35-mm slide,
6 × 9.25' = 55.5' to view the 35-mm widescreen, and
6 × 13.3' = 79.8' to view the 16-mm cinemascope.

But the viewer is looking at the same graphics in each of the above instances—so if the first distance of 39.9 ft is correct (which it is, because the 6W rule was based on an image width equal to 1.33 times the image height), then the other distances are not valid. A moment's reflection by the reader will reveal that it is the *height,* not the *width* of the image, that accounts for the readability of the captions. Certainly the letters A, B, C, of the same size and spacing in each of the 5-ft high images, do not care how *wide* the image format is.

Using H then, *not* W, as the criterion for readability of standard graphics, we can determine that 6W is the same as 8H for the 16-mm format. Thus 6 × 6.65 ft = 40 ft, and 8 × 5 ft = 40 ft. We can now write a revised rule that says: The height H in Fig. 1-9 of a projected image should be equal to ⅛ of the maximum viewing distance V, measured from side of screen, diagonally to far seat. Also, the farthest viewer should not sit more than 8 image heights from the screen. This is the meaning of column (4) in Table 1-1. Note that certain entries in this column show V ÷ 5.4 instead of V ÷ 8. This results from taking the 35-mm slide, and turning it vertically. Caption characters would still be the same size as when the slide is horizontal, but there now are *more* characters to fit into the new vertical height. Viewing distance is, of course, unchanged, so that it only requires 5.4H to equal 8H of the horizontal 16-mm format. If the 35-mm slide image is 5 ft 0 in high, its width is 7.4 ft as previously calculated, when used vertically, the image is 7.4 ft high, and 7.4 × 5.4 ft = 40 ft, as before.

The superslide and all other square formats are projected at the same height as the vertical 35-mm slide, and therefore the 5.4 factor applies. The variable 6 to 8 factor shown in Table 1-1 for cinemascope, wide screen, and 70-mm permits the showing of larger-than-needed images for more impact in theater installations. Thus, a theater designed for 6H viewing, instead of 8H, assures extra large images.

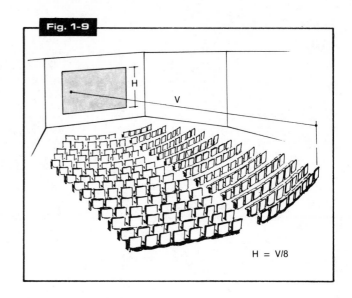

Fig. 1-9

H = V/8

CEILING TOO LOW

SCREEN SIZE VERSUS CEILING HEIGHT

Once we have become aware of the proper image sizes needed for good viewing, we are more likely to be conscious of the importance of ceiling height in designing AV (audio visual) spaces. The problem of low ceiling height depicted in Fig. 2-1 is getting to be more prevalent, with the many large corporations, banks, insurance companies, and so forth, which are occupying multistory office buildings.

Most of these enterprises have need for a large meeting room and a corporate board room, as well as numerous conference rooms, training rooms, and specialized management rooms. Many have a 200- or 300-seat auditorium in their planning, also. Large meeting spaces are especially likely to suffer from low ceiling-height restrictions.

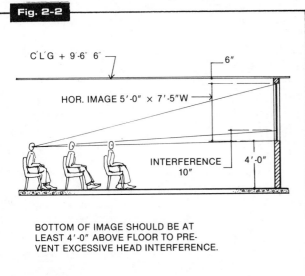

BOTTOM OF IMAGE SHOULD BE AT LEAST 4'-0" ABOVE FLOOR TO PREVENT EXCESSIVE HEAD INTERFERENCE.

WHAT IS THE PRACTICAL SCREEN SIZE RESULTING FROM THE STRUCTURE LIMITATION OF A 9 FT-6 IN CEILING HEIGHT?

The 9 ft-6 in ceiling height is so common in commercial buildings that its limitations should be thoroughly understood. It is worthwhile to consider the maximum length of room that can be served by a projected image in such a space. In Fig. 2-2, we see that a seated viewer will experience about 10 in of head interference when seated on a flat floor under the following conditions:

- rows have staggered seats, so it's the person's head two rows in front of the viewer that interferes;
- rows are no closer than 36 in, chair-back to chair-back;
- bottom edge of image is 4 ft-0 in above the finished floor.

Fig. 2-2 shows that the height of the image is limited to 5 ft-0 in, allowing 6 inches at the top of the image for proper aesthetic appearance. With an aspect ratio of 1.48:1 (Table 1-1, Section 1), the maximum 35-mm horizontal slide image size is 5 ft-0 in high × 7 ft-5 in (5.0 ft × 1.48 = 7.4 ft). This is the maximum usable screen size.

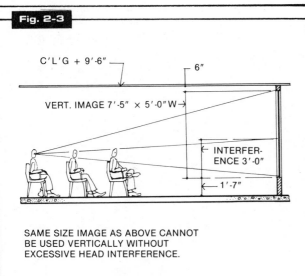

SAME SIZE IMAGE AS ABOVE CANNOT BE USED VERTICALLY WITHOUT EXCESSIVE HEAD INTERFERENCE.

CAN VERTICAL SLIDES BE USED WITH THE ABOVE SCREEN SIZE?

No. When vertical slides are used, they have the same dimensions as horizontal slides, only they are turned upright (Fig. 2-3). This means that the 7-ft-5-in dimension will be vertical, and the 5-ft-0-in dimension, horizontal. Fig. 2-3 shows that now with the slide oriented vertically, the bottom of the image is only 1 ft-7 in above the floor. This produces head and shoulder interference, obliterating half the image height, which is unsatisfactory. Further, if vertical and horizontal slides were intermixed in the slide tray, the horizontal slide centerline would drop to the center of the 7-ft-5-in dimension, thus lowering the bottom edge of the horizontal image to 2 ft-9½ in above the floor. This is 1 ft-3 in less than the 4-ft-0-in recommended height and is, therefore, unsatisfactory, also.

If vertical slides are to be used properly, then the ceiling height must be 12 ft-0 in.

Exercise:

SKETCH A FRONT VIEW OF THE DUAL-IMAGE SCREEN OF MAXIMUM PRACTICAL SIZE FOR A ROOM WITH A 9-FT-6-IN CEILING, SHOWING 16-MM FILM IMAGE, SINGLE CENTERED SLIDE, AND DUAL SLIDE, HORIZONTAL FORMAT. ALLOW 6 IN ABOVE TOP OF IMAGE. (Your sketch should look something like Fig. 2-4.)

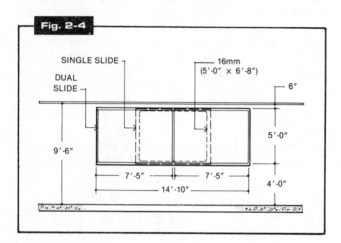

SKETCH SCREEN CONFIGURATION FOR DUAL-IMAGE HORIZONTAL AND VERTICAL SLIDES, SAME SIZE AS ABOVE. LET TOP OF IMAGE CLEAR CEILING BY 3 IN. COMMENT ON THIS ARRANGEMENT. (Refer to Fig. 2-5.)

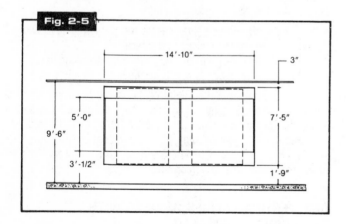

Comment:

Inasmuch as vertical and horizontal slides are the same size (turned 90°), the height of the screen must be equal to the width of the horizontal slide. In other words, the screen is twice as wide as it is high, for dual image. Unfortunately, when the long slide dimension is placed vertically, there remains insufficient space below the vertical image, even when the top edge of the image is raised to within 3 inches of the ceiling.

Note that in Fig. 2-5 the horizontal images drop down so that their lower edge is only 3 ft-½ in above the finished floor. This is so because the common center for both formats must be located at the center of the governing vertical slide.

The important point to learn from this analysis is that vertical slides cannot be considered in structures where the ceiling height is limited to the common 9 ft-6 in dimension, unless the vertical slide has a maximum height not exceeding 5 ft-0 in. When this slide is turned horizontally its height is 5 ÷ 1.48 = 3.38 feet (see Table 1-1), and the maximum viewing distance is 8 × 3.38 ft = 27 ft. We arrive at the same answer by using the vertical height of 5 ft-0 in and multiplying it by 5.4 (Table 1-1).

When we confine the projection display to horizontal slides only, the image height can be a maximum of 5 ft-0 in, and hence the maximum viewing distance can be 8 × 5 ft = 40 ft, with an image 5 ft-0 in high by 7 ft-5 in wide.

DETERMINING THE CEILING HEIGHT REQUIRED FOR VARIOUS VIEWING DISTANCES

The Chart 2-1 considers two cases, i.e., Case I, where either movies or slides are shown in horizontal format, and Case II, where vertical and horizontal slides are to be projected. In Case II, the fact that the screen is as high as it is wide permits the use of superslides which have a square format.

Dual image screens may of course be used by adding another image section. If this were done in Case I, the final screen size would be 2 × 1.48 = 2.96 times as long as it is high, and for Case II, the screen would always be twice as long as it is high. When dual images are used, 16-mm images are centered.

Example:

The distance measured diagonally from one corner of the screen to the farthest viewer (estimated) is 45 ft. No vertical images will be projected. Find the required ceiling height.

Solution:

From Chart 2-1, the Case I ceiling height is 10 ft-2 in, say 10 ft.

CHART 2-1

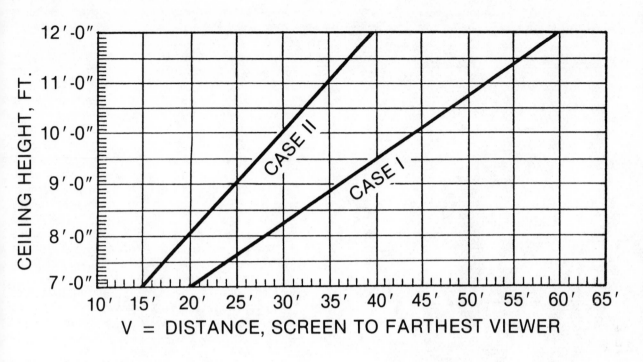

V = DISTANCE, SCREEN TO FARTHEST VIEWER

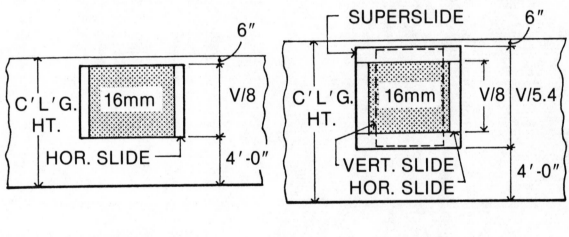

CASE I

CASE II

17

There has been a good deal of controversy over the years concerning the relative merits of front versus rear screen projection. It is not true that one is inherently better than the other. It is true only that one is more suited for a certain application than the other. Once we understand the characteristics and limitations of each, the proper application becomes obvious.

ON WHAT PRINCIPLE DOES THE FRONT PROJECTION SCREEN DEPEND?

The principle of *reflection* is illustrated in Fig. 3-1. If it is imagined that the screen is a mirror, then we could say that the rays of light leave the projector lens, travel in straight lines, and impinge on the screen surface from which they are reflected into the audience area.

A true mirror surface would cause the reflected rays to return in an ever widening pattern, inasmuch as their angles of incidence must equal their angles of reflectance. Actual screens are, of course, not mirrors. If they were, the viewer would see no image, except the bright light of the projection lamp shining in his eyes by reflection. Although not mirrors, the screens *are* reflectors, but the reflection is *diffuse*, not *specular*. Nevertheless, the mirror concept is helpful for analyzing how the screen *tends* to work, depending on how specular or diffuse its surface is.

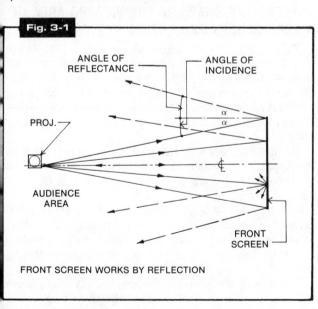

FRONT SCREEN WORKS BY REFLECTION

ON WHAT PRINCIPLE DOES THE REAR PROJECTION SCREEN DEPEND?

The principle of *transmission* is shown in Fig. 3-2. If it is imagined that the screen is a transparent sheet of glass or plastic, then we could say that the rays of light leave the projector lens, travel in straight lines, and impinge on the screen

FRONT VERSUS REAR PROJECTION

surface through which they are transmitted into the audience area.

A truly transparent screen would permit the rays to travel, undeviated, with an ever-widening pattern quite similar to that just described for the front screen. The difference lies in the wider spread of the rays leaving the lens of the rear screen projector, necessitated by the short throw distance used in practice to minimize the depth of the projection room.

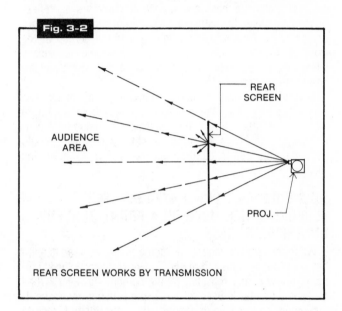

REAR SCREEN WORKS BY TRANSMISSION

This dissimilarity could pass unnoticed, but, as we will learn later, it has an important bearing on the shape of the good viewing area.

Just as the front screen is never truly reflective, the rear screen is likewise never truly transparent. If it were, again we would be looking right into the projection lamp imaged in the lens. The effect would be like looking into an automobile headlight. But again, the transparent screen concept is useful for analyzing how the screen *tends* to work, depending on how diffuse the surface is.

WHAT IS AMBIENT LIGHT?

Ambient light refers to any light, other than the light from the projection beams, that is visible in the area under discussion. Training rooms, lecture halls, and conference rooms operate in a rather high ambient light so that the participants, as well as the presenter, can see each other and see to take notes or read text.

On the other hand, theaters, auditoriums, and multimedia display rooms, where the only audience involvement is emotional, are more suited to a relatively dark ambient lighting environment.

WHY IS THE REAR PROJECTION SCREEN WELL SUITED TO COMBAT AMBIENT LIGHT?

Rear projection screens perform exceptionally well in lighted rooms because most of the ambient light that strikes the screen is *transmitted* through the screen into the projection room, where it is absorbed in the dark, nonreflective surfaces of the walls and furnishings.

HOW DOES THE FRONT PROJECTION SCREEN HANDLE AMBIENT LIGHT?

All front projection screens work by reflected light, consequently, any ambient light that can be "seen" by the screen will be reflected into the audience area. Thus, any viewer seated in the path of such reflected ambient light will experience a degrading of the image. If the reflected ambient light is strong enough, the image can be completely "washed out." It is obvious that room lights which permit stray reflections to illuminate the screen are a matter of concern to the AV designer. Fluorescent ceiling lights, near the screen at the front of the room, are likely to be troublesome, especially when they are equipped with the standard office-type crystalline plastic lenses.

Several manufacturers make light-controlling lenses and accessories to solve the problem, ranging from egg-crate louvers to parabolic grids; but regardless of these light control devices, the rows of lights nearest the screen wall should be separately controlled from the remaining lights in the room.

HOW DOES AN ILLUMINATED POINTER PERFORM WHEN USED ON A FRONT SCREEN? ON A REAR SCREEN?

When the familiar hand-held electric pointer is used to point out subject matter on a front screen, we see clearly the illuminated arrow or dot reflected by the screen. However, when this same pointer is used on a rear projection screen, the illuminated marker fails to appear with sufficient brilliance to be distinguished from the background of the projected image. The reason for this, of course, is that the rear screen *transmits* the incident light, and does not reflect enough of it to be seen by the audience. The illuminated arrow or dot will appear on the *back* side of the screen.

The reflectance of a rear projection screen is about 0.18, which means that it reflects 18% of the ambient light falling on it from the audience side. Consequently, if a pointer is to be seen by reflection, it must be bright enough so that even with only 18% of its light reflected, it is easily visible to the audience.

To fulfill these requirements, a laser beam light pointer has been developed. It produces an intense beam of red light, forming a ½-in diameter spot on the screen that can be seen from the back of the room, even though 82% of the light travels right through the screen!

WHAT IS THE SHAPE OF THE VIEWING AREA IN A FRONT PROJECTION SYSTEM?

The front screen viewing area is basically sector-shaped. The farthest boundary is formed by two 8H arcs that cross at the centerline of the screen, and have their centers at the diagonally opposite sides of the screen as diagrammed in Fig.

3-3. The near boundary is formed by two lines displaced 60 from the undeviated rays reflected from each side of th screen. It should be noted that this undeviated ray is the per fect reflection of a projection ray from the lens to the edge o the screen, where the angle of incidence is equal to the angl of reflection.

The 60° represents the practical maximum "bend angle," or angle of diffusion, through which the screen is able t spread the light into the eyes of the audience.

A front-row boundary line 3H (2½H min.) from the scree defines the location of the first row of seats, and when full sector seating is not used, boundaries formed by the sid walls are imposed.

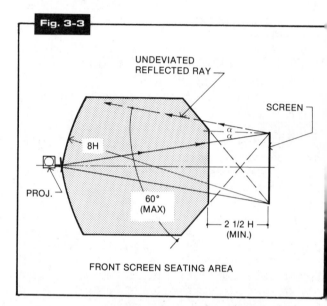

FRONT SCREEN SEATING AREA

WHAT IS THE SHAPE OF THE VIEWING AREA IN A REAR-SCREEN PROJECTION SYSTEM?

The viewing area for the rear-screen projection system (Fig. 3-4) is also sector shaped, but its angular spread is not as great as that of the front-screen pattern. This is because the maximum bend angle is usually less than 60°, 55° being a workable value. Note that the undeviated ray passes *through* the edge of the screen by transmission, and not by reflection.

The rear boundary arcs are again formed by 8H radii from the diagonally opposite sides of the screen, and the distance from screen to front row of seats is again taken as 3H (2½H min.).

Note that the limiting dimensions in the preceding two analyses are based on H, the height of the image, and not W, its width. The reader should review Section 1 for the reasoning behind the use of H as the more meaningful measuring unit.

It must be remembered that the bend angle is nothing more than the ability of a given screen material and surface to bend, or diffuse, the projected light rays from the lens so that they can be seen by a viewer seated on the opposite side of the room. The implication is that if a viewer can see a brilliant enough image coming from the far side of the screen, from the viewpoint on the near side of the screen, then images can be seen brilliantly coming from all intermediate points on the screen.

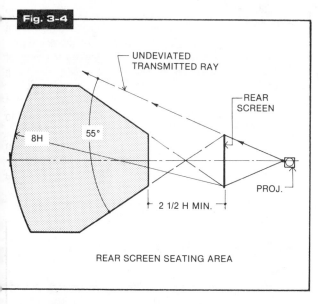

Fig. 3-4

UNDEVIATED TRANSMITTED RAY

REAR SCREEN

8H

55°

2 1/2 H MIN.

PROJ.

REAR SCREEN SEATING AREA

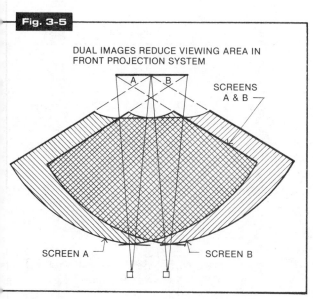

Fig. 3-5

DUAL IMAGES REDUCE VIEWING AREA IN FRONT PROJECTION SYSTEM

A   B

SCREENS A & B

SCREEN A

SCREEN B

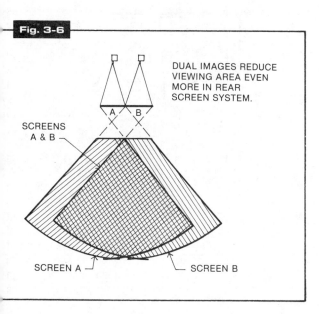

Fig. 3-6

DUAL IMAGES REDUCE VIEWING AREA EVEN MORE IN REAR SCREEN SYSTEM.

SCREENS A & B

A   B

SCREEN A

SCREEN B

In comparing a standard type of rear screen with a front screen, it can generally be said that the rear screen is not capable of diffusing, or scattering, light as efficiently as the front screen. Consequently, we should be prepared to accept a more restricted audience area when designing with rear-projection screens. There are some exceptions to this statement, and these will be discussed further in Section 4.

WHAT HAPPENS TO THE VIEWING AREA WHEN DUAL IMAGES ARE USED?

Front Projection

When images are projected side-by-side, in dual format, each image has its own viewing area, as shown in Fig. 3-5. When the two viewing areas are superimposed, one on top of the other, a portion of each will remain common to *both* images. This portion appears with double crosshatch in the illustration, and represents the area within which viewers can see both images with satisfactory brilliance. A viewer seated to the left of this common area will see the left screen with good brilliance, but the far right portion of the adjacent image will exhibit an appreciable light fall-off.

Rear Projection

The above remarks hold true for rear-screen projection, also. Note, however, that the individual viewing areas for images A and B are smaller than those for the front projection. Consequently, as Fig. 3-6 shows, the common area is smaller for images A and B.

The reason, of course, for this reduction in area is twofold: (1) the undeviated ray angle, from which the 55° diffusion angle is measured, is spread more widely due to the wide-angle lens necessitated by the short-throw distance, and (2) the diffusion angle itself is only 55°, due to the characteristics inherent in transmission-type screens.

ARE TRIPLE IMAGES DESIRABLE?

Generally, no. Viewing areas are further reduced, resulting in much space being wasted in the front part of the room. The exception is when a multimedia display is being used, in which case screen brilliance for all seats, for all images, is secondary.

WHAT IS THE PRINCIPLE OF OPERATION OF THE CURVED SCREEN?

The curved screen, like the flat front screen, works by reflection. The curvature, shown in Fig. 3-7, directs the undeviated reflected rays into the audience area where they meet at a point called the *optimum seat*. A viewer seated here will see the entire screen with equal brightness.

Diffused rays, scattered up to 40° from the undeviated ray, will cover a wider seating area than those of the flat screen with a 50° scatter from the undeviated reflected ray. This is due entirely to the curvature of the screen. Also, rays scattered up to 20° from the undeviated ray will cover 80% of the audience area, which means that most of the audience sees a brilliant image.

COMPARE THE SEATING AREA OF FRONT FLAT, FRONT CURVED, AND REAR FLAT SCREENS

Refer to Fig. 3-8. The illustration shows the full-sector seating area for each of the three types of screens. Normally the full-sector area is not used, so the dashed lines indicate the superimposition of a rectangular room, with side aisles.

If we let the area so formed for the curved screen equal 100%, then the flat front-screen area represents 88%, and the flat rear screen, 82%. These percentages, while representative of actual screens, may vary depending on the kind of screen material used. For example, some rear screen materials may produce only a 69% audience area compared to that of the curved screen. Care should be taken, when specifying screens, to be sure that the screen characteristics are compatible with the required performance. Section 4 discusses screen characteristics further.

Table 3-1 lists some recommended uses of front, rear, and curved screens.

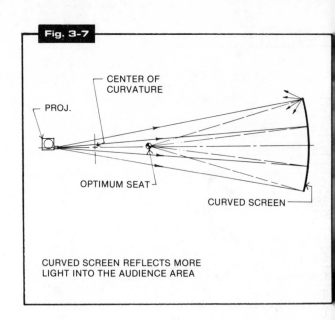

Fig. 3-7

CURVED SCREEN REFLECTS MORE LIGHT INTO THE AUDIENCE AREA

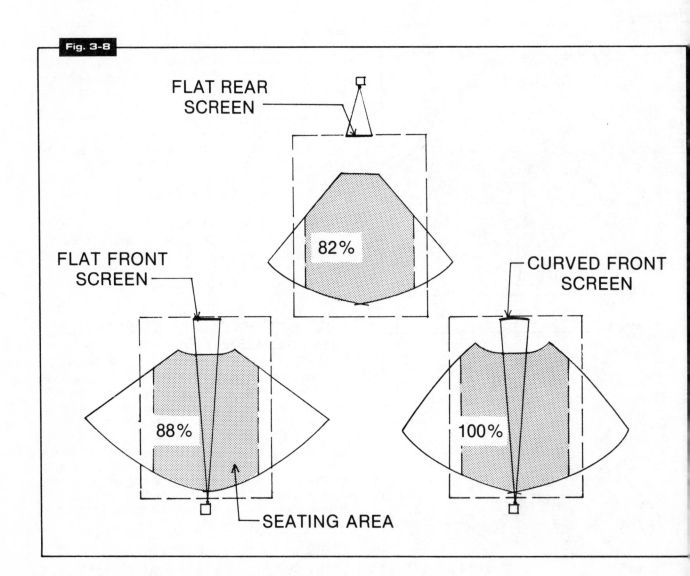

Fig. 3-8

Table 3-1. Recommended Uses of Front, Rear, and Curved Screens

Application	Type of Screen		
	Flat Front	Flat Rear	Curved Front
Auditorium (up to 500 seats)	✔	✔	✔
Auditorium (500-1500 seats)	✔	✔	✔
Auditorium (1500-4500 seats)	✔		✔
Ballroom	✔		✔
Convention Hall	✔		✔
Dining Area	✔		
Theater	✔		✔
Lecture Hall	✔	✔	
Church	✔		
Classroom	✔	✔	
Training Room		✔	
Conference Room	✔		
Multimedia Room	✔	✔	✔
Boardroom	✔	✔	✔
Meeting Room (Large)	✔		✔
Management Information Room	✔	✔	
Outdoor Movies	✔		✔
Display-Lobbies, Public Spaces	✔	✔	
Museums, Fairs	✔	✔	✔
Screening Rooms	✔		✔

PROJECTION SCREEN PERFORMANCE

The projection screen is an important part of the projector-screen-viewer system but, unfortunately, it is an item that is often taken for granted. The specifying of a projection screen should be done with as much thought as selecting the projector itself. There are many kinds of projection screens, both front and rear, and each kind has its own characteristics, which may be chosen to satisfy a given application.

Rear projection screens are particularly vulnerable to misapplication and, consequently, should be very carefully selected. Front screens, on the other hand, are less critical in their performance and the chances of misapplication are not so likely. The discussion in this section will explain the basic characteristics of projection screens and how they affect performance.

WHAT IS THE MAIN PURPOSE OF THE TEST PERFORMED ON A REAR SCREEN?

The main purpose in testing a rear projection screen is to investigate its light-diffusing ability. Screens that transmit the projected light beam without sufficiently diffusing it, or scattering it, to cover the audience area, are known as "hot" screens. Hot screens exhibit a well-defined hot-spot wherever a viewer's eye is in line with the lens of the projector. The hot-spot therefore will move across the screen as a viewer moves from one side of the room to the other within the extended beam angle of the projected rays of light. The hot spot is formed because the eye is looking into the lens, at the lamp filament, right through the screen material. In Fig. 4-1, point C represents the observer's hot spot on the screen, and point D is a typical dim spot.

The enlarged view at the right in the figure shows why there is such a variance in screen brightness from a given observer's viewpoint. It is clear that the screen has a difficult job to bend the transmitted ray of light at point D through a *bend angle* of 65° along path D-D′ into the eye of the observer. By comparison, the undeviated ray C-C′ encounters no bend angle in reaching the observer's eye, and thus is at maximum brightness for the given screen material.

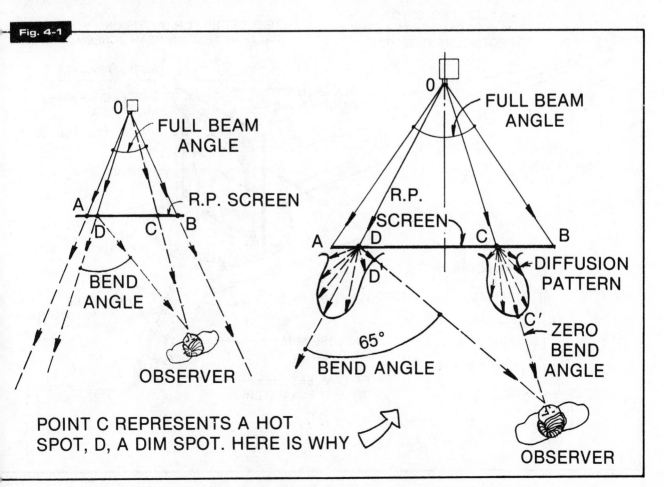

Fig. 4-1

FULL BEAM ANGLE

R.P. SCREEN

BEND ANGLE

OBSERVER

POINT C REPRESENTS A HOT SPOT, D, A DIM SPOT. HERE IS WHY

O FULL BEAM ANGLE

R.P. SCREEN

65° BEND ANGLE

DIFFUSION PATTERN

ZERO BEND ANGLE

OBSERVER

The light diffusing ability of the screen is represented by the diffusion patterns shown in Fig. 4-1. These patterns appear as cusps, each outlined by an "envelope" formed by dozens of little vectors representing scattered rays, whose lengths are proportional to their brightness.

DESCRIBE A DIFFUSION PATTERN TEST ON A REAR SCREEN SAMPLE

In a typical diffusion pattern test on a sample of rear screen material, the screen sample is mounted with its dull, or coated, side, facing away from the light source which will be used to illuminate it, and the transmitted brightness of the illuminated spot of light is read with a footlambert meter, while traversing the spot in a 180° arc. Brightness readings are taken every 10° for each of three angles of incidence of the spot light-beam. The angles of incidence correspond to the angle at which a projected light-beam strikes the screen when it is illuminating the left third, center third, and right third of the projected image, as shown in Fig. 4-2.

The center points of the left, center, and right thirds of the screen image have been chosen to represent typical points of observation by a viewer as he or she scans the screen for important information. These points have been labeled L, C, and R, in Fig. 4-2, and the angle of the incident ray at these particular points is 8°, 0°, and 8°, respectively. Inasmuch as

these angular orientations are simulated in the test, som means of rotating the dial and wood frame assembly inde pendently from the traverse arm must be provided. The di and frame assembly can then be set at an 8° angle of inc dence relative to the light spot beam, on both sides of the cen ter position, thus simulating the actual projected ray angle a points L, C, and R.

The official name of an instrument that performs this kind a test is *photogoniometer*, or *goniophotometer*. Such a instrument, commercially obtainable, is sophisticated an costly. However, the instrument shown in Fig. 4-2 was mad in a home workshop, at a total cost of about $12.00, exclud ing the brightness footlambert meter. The light source is standard 35-mm slide projector, using a 2- × 2-in met slide, with a ¼-in diameter hole drilled on-center. When th projector is located 13 ft-4 in from the test screen, it will pro ject a 10-in diameter light-beam onto the screen sample Screen samples smaller than 12 in square are not as easi tested, and should be avoided.

In operation, the light beam and screen sample are set the correct angle of incidence for the *actual* lens that will b used, for each point of test L, C, and R. A *footcandle* mete held with its back against the screen, in the center of the sp and facing the projector, reads the incident light reaching th screen.

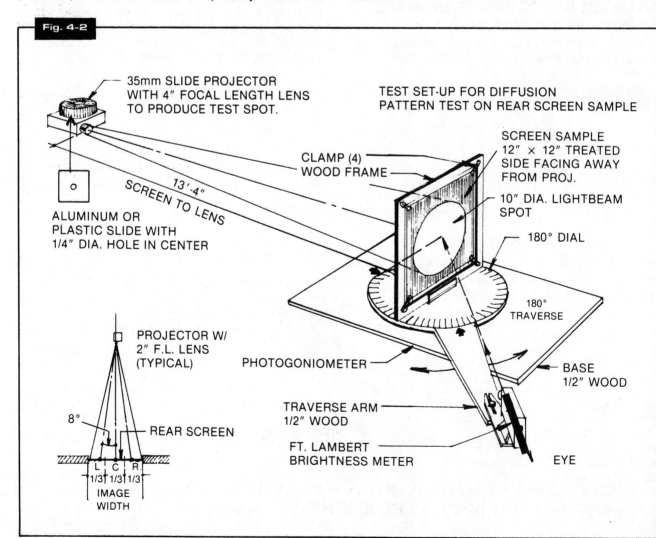

Fig. 4-2

35mm SLIDE PROJECTOR WITH 4" FOCAL LENGTH LENS TO PRODUCE TEST SPOT.

TEST SET-UP FOR DIFFUSION PATTERN TEST ON REAR SCREEN SAMPLE

SCREEN TO LENS 13'-4"

ALUMINUM OR PLASTIC SLIDE WITH 1/4" DIA. HOLE IN CENTER

CLAMP (4)
WOOD FRAME

SCREEN SAMPLE 12" × 12" TREATED SIDE FACING AWAY FROM PROJ.

10" DIA. LIGHTBEAM SPOT

180° DIAL

180° TRAVERSE

PROJECTOR W/ 2" F.L. LENS (TYPICAL)

PHOTOGONIOMETER

BASE 1/2" WOOD

8°

REAR SCREEN

TRAVERSE ARM 1/2" WOOD

FT. LAMBERT BRIGHTNESS METER

EYE

L C R
1/3 1/3 1/3
IMAGE WIDTH

Note that incident light is measured in footcandles. When this spot of light is viewed from the front, or "audience side" of the screen sample, its brightness will depend on three things: the light-transmitting characteristics of the screen material and its coating, the angle of view, and the amount of ambient light which the screen faces. If the test is run in a darkened room, the last item can be ignored. Whereas the incident light from the projector is measured in footcandles, the transmitted light, evidenced as brightness to the eye, is measured in footlamberts. It is evident that the footlambert is nothing more than a footcandle after it has been acted on by the screen material. Such action may result in a footlambert value less than, equal to, or more than the original footcandle value of the incident light.

WHAT IS THE FUNCTION OF THE COATING APPLIED TO THE SCREEN?

The purpose of the coating applied to the screen is to disperse the light, or spread it into the viewing area. In doing this, the coating acts like millions of tiny lenses and, thus, certain coating formulations actually can magnify the light in controllable directions, at the expense of a diminishing of the light in others. The more a screen tends to magnify the light seen by an observer along an undeviated ray, the less brilliant the screen will appear as the observer scans an area of the screen displaced from his original line of sight.

In a typical application, long narrow rooms can utilize a screen formulation that magnifies the projected light in a narrow beam. Thus it is possible to use a projector with a lower light output and still serve most of the audience area with suitable screen brilliance. Such a screen diffusion pattern is shown in Fig. 4-3A. Note again that it is sufficient to take test readings at three places on the horizontal centerline of the screen, marked L, C, and R. Also note that the angle of the incident ray at L and R is 8° off center. This particular angle will result whenever a lens of 2-in focal length is being used in the actual system. Screen coating reaction to incident light is affected by the angle of incidence. The greater the angle of incidence, the more the brightness drops off at the sides of the screen. For this reason, it is well to avoid lenses with focal lengths less than 2 in.

WHY WAS A 4-IN FOCAL LENGTH LENS USED IN THE TEST SET-UP, WHILE THE TESTS SHOW DIFFUSION PATTERNS FOR A 2-IN LENS?

The 4-in lens used in the test set-up has nothing to do with the fact that the diffusion patterns are plotted for a 2-in lens. The purpose of the 4-in lens was only to produce a spot of light at a very narrow cone angle, to simulate the collimated (parallel light rays) light-beam of an expensive photogoniometer. This small spot of light was then aimed at an angle relative to the test screen, equal to the actual angle of incidence occurring with the designated system, at points L, C, and R. What we are doing is seeing how the sample screen transmits a given intensity (the footcandle reading taken on the incident side of the screen sample) of light, and how the resulting brightness appears on the audience side of the sample (the footlambert brightness reading over the 180° scan). Once the relationship between the incident light and the transmitted light is established, the diffusion pattern can be plotted, as shown in Fig. 4-3A, B, and C.

HOW ARE THE RESULTS OF SUCH A RELATIONSHIP INTERPRETED?

A quantity to be called the *figure of merit,* symbolized by "Q" in Fig. 4-3C, has been used to express the relationship between the light intensity received by the screen, at a given point, and the brightness observed at various viewing angles about that point. Thus, if the screen shown in Fig. 4-3B received 23 footcandles at point L, from the test projector with the 4-in lens, and if the maximum brightness read along the 8° undeviated ray line was 28 footlamberts, then at this point and angle of view, the screen is performing with a Q = 28 ÷ 23 = 1.22. This means that whatever the actual light intensity is for the given system, the screen will transmit 1.22 times as much light to an observer seated along the 8° ray line extended into the audience. Should he look to the right, at point R on the screen, his line of sight might intersect the diffusion pattern at a Q value of 0.3, indicating that only ⅓ of the light received by the screen is transmitted as brightness along that particular sightline. It is thus possible, with the above type of diffusion patterns, to determine how bright the principal points L, C, and R on the screen will appear to a viewer located anywhere in the viewing area.

It should be noted in Fig. 4-3 that the Q values, indicated by little vectors within the diffusion patterns, are not plots of the brightness value at the given angle, but are plots of the *ratios* of the light received divided by the light transmitted.

WHAT SHAPE ROOM CAN USE A REAR SCREEN WITH DIFFUSION PATTERNS LIKE THOSE SHOWN IN FIG. 4-3B AND C?

Rooms whose length-to-width ratio is about 1.25 to 1.5 can profit from the use of screens that provide more screen brightness at the extreme side-viewing angles. The screen shown in Fig. 4-3B is an example of such a screen. Its maximum Q is only slightly greater than that of test A, but its ability to spread the light at the medium wide angles of viewing is appreciably improved. It still does not do noticeably better than A at the very wide angles of viewing.

Screen C, however, shows noticeable improvement over both A and B in Fig. 4-3. While its use is definitely called for in square or wider-than-long rooms, it would naturally enhance the image in any shape room. Its maximum Q is about 1.55. The interesting thing in comparing the tests on these three screens (all taken from actual screen samples of formulations currently in use and widely advertised) is that they are all priced the same! Of the three, obviously screen C is by far the best choice.

HAVE THERE BEEN ANY RECENT NOTEWORTHY DEVELOPMENTS IN REAR SCREEN DESIGN?

For the most part, rear screens using glass or acrylic substrate material have had improvements concerned with the formulation of the light-diffusing coating applied to their surface. This goes for the flexible plastic rear screens also. Unless some new light diffusing coating is discovered, these types of screens have probably reached their peak of performance, typified by screens represented by test C, Fig. 4-3C.

Further improvements seem to be possible with the acrylic substrate type, however, because it can be molded to comprise various elements of the well-known Fresnel lens

Fig. 4-3

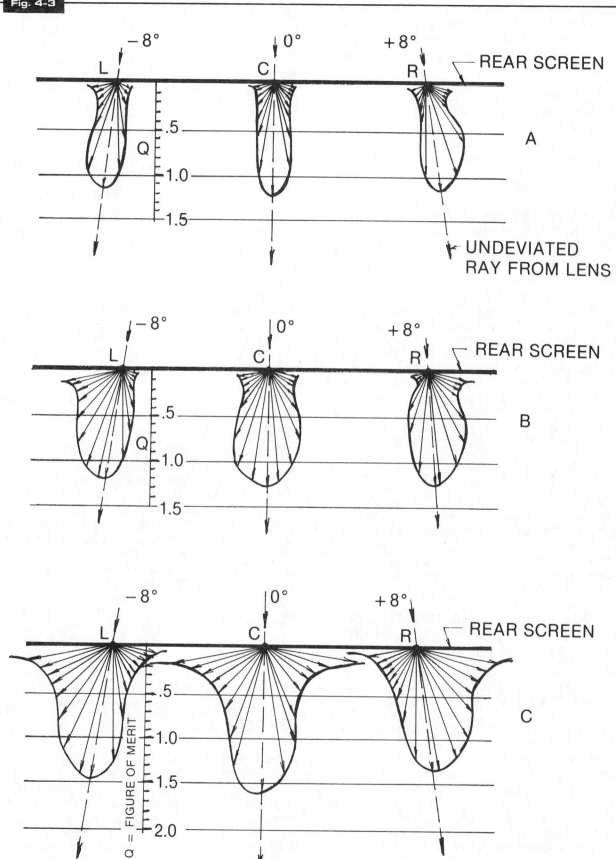

inciple. It is thus possible to capture a good part of the light
at gets diffused but never reaches the eye, because it
avels upward or downward from the average plane of view-
g. If enough of this stray light could be captured and redi-
cted into the eye zone, marked increase in brightness
ould result.

Acrylic screens using this principle are now being tested,
d one manufacturer has already marketed one such device,
hich has such an abundance of brilliance that it can afford
ack vertical stripes, equally spaced and covering 50% of its
ont surface, and still out-perform the plain surface type of
creen. The 50% black stripe area obviously eliminates 50%
 the room ambient light reflection, so that the screen can
ork in twice as much ambient light as the plain surface
creens. The black stripes, measuring 1/10 in wide, are
visible from a distance of 10 to 12 ft and, therefore, do not
terfere with the image as seen by the eye.

A reproduction of the actual test plot on this high-per-
rmance screen is shown in Fig. 4-4, plotted to the same
cale as the plots A, B, and C in Fig. 4-3, for comparison.
sadvantages? There are two: high cost and a complicated
ssembly procedure. But, if the performance of this screen is
y indication of what we can expect in the near future, then

the AV industry will at last be able to fill one of the gaps that
has been sorely missing in the last 25 years!

WE HAVE TALKED ABOUT FIGURE OF MERIT. WHAT ABOUT SCREEN GAIN?

Figure of merit and screen gain are terms that have very
similar usage. In the early days of front projection, it was
known that a projection screen that had a flat, white, matte
surface reflected the projected light beam in an extremely
uniform manner at all viewing angles. The only problem was
that as audience areas increased, larger images were required, but projector light sources (other than arc lamps in
theater projection) were found short on light output. Ways
were found to make front projection screens more reflective
and directional. The problem was always the same—gather
the light and aim it down the center of the house. Aluminized
surfaces, pearlescent surfaces, and lenticular (embossed)
surfaces, to name a few, were used to ''magnify'' the reflected light for the central prime-seating area, at the expense
of the extreme side-aisle seats.

Screens that increased the light in certain directions had a
''gain'' factor, and this gain became an important screen
characteristic. The obvious way to measure gain was to com-

Fig. 4-4

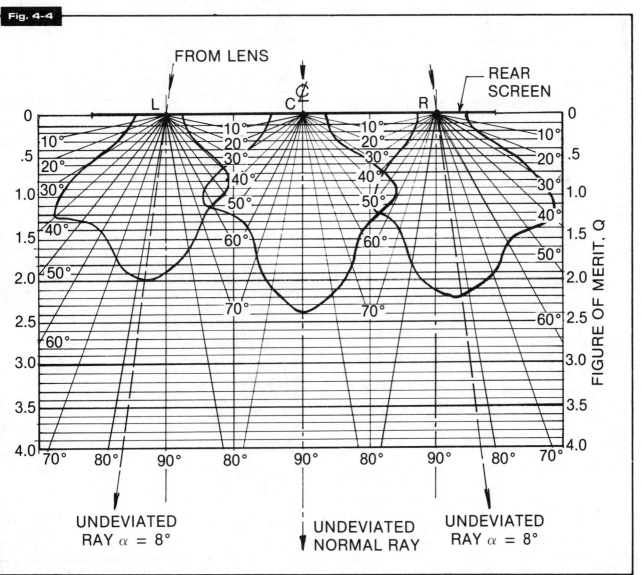

pare the light reflected by a perfect matte surface with the light reflected by the surface under test. The nearest thing to an absolutely nonreflective, uniform-scatter surface was found to be the flat surface of a block of magnesium carbonate, a white chalky substance. This, then, became the standard of comparison for all screens with gain characteristics.

In analyzing the performance of a given projection system, however, with a given screen, projector, lens, light source, and audience area, it seemed more logical to analyze the screen performance by comparing the amount of light received by the screen to the brightness produced by the screen, as seen by observers in selected locations throughout the audience area. The author has therefore introduced the use of a figure of merit, Q, to express this relationship. The figure of merit is system-oriented, and straightforward in use. It is simply a ratio, and therefore dimensionless.

WHAT DO WE MEAN WHEN WE SAY THAT Q IS "SYSTEM ORIENTED"?

The word system here refers to an actual installation where we have a projector, screen, and viewer. This is a system. In evaluating a system like this, we can see that it makes sense to examine the actual light output of the projector, couple it with the screen characteristics, and determine the actual viewer screen brightness from various viewer locations.

HOW DO WE MAKE THE ACTUAL SYSTEM ANALYSIS?

The technique for making such an analysis is as follows:

1. Make a scale layout of the room, usually $\frac{1}{4}$ in = 1 ft-0 in is satisfactory.
2. Reproduce the diffusion patterns for the three screen target points L, C, and R (such as shown in Fig. 4-4) to the same scale as the room layout, in their correct position against the screen.
3. Plot a light-distribution curve for the projector, and draw it adjacent to the back side of the screen, on the scale drawing, as illustrated in the Fig. 4-5.
4. Divide the audience area into convenient zones, and add typical viewing positions as desired.
5. From each observer position draw three lines to target locations L, C, and R, and read Q at intersection with diffusion curves.
6. Calculate brightness at L, C, and R as seen by the viewer, for all viewer locations chosen.
7. Calculate ratio of brightest spot to dimmest spot (logarithmically, to simulate the way the eye perceives brightness), and determine which seats satisfy the accepted ratio of 3:1. This will determine the optimum seating area for good viewing.

Steps 1 and 2 are easily accomplished. Step 3 needs some explanation. To plot a light-distribution curve, divide the actual image width into 10 divisions, making 11 positions, counting the 2 extreme positions. Take a footcandle reading across the horizontal center of the screen at each position, as shown in Fig. 4-6. Use a clear mask opening in the gate, with no glass, and no transparency. If you cannot take readings at the site, use a similar set-up in the shop, with the same lens and light source.

If readings are not symmetrical (see Fig. 4-6), average the

left and right corresponding locations. On an arbitrary grid, drawn at the back side of the screen, with the vertical scale representing 10 divisions, or 100% (1.0), plot each reading as a percentage of the maximum center reading.

Note: If the center reading is not greater than the others, then the lamp is not centered properly. Corresponding readings should be symmetrical within 10%, ideally.

In Fig. 4-6, the center reading, 29 footcandles, is plotted as 100%, or 1.0. The next reading is plotted on each side of center as

$$28 \div 29 = 0.97 \text{ or } 97\%$$

The next set, as

$$\frac{27 + 26}{2} \div 29 = 0.91 \text{ or } 91\%, \text{ etc.}$$

It will be found that the end readings, 15 and 16, are right on the edge of the image format, and cannot be read unless the slide mask is removed, and the readings taken at the spot where the mask edge was imaged.

Steps 4 and 5 are explained in the illustration.

Step 6 shows, that in Fig. 4-6, the target point L receives 80% (0.8) of the center illumination (29FC), and the screen alters this by a factor Q = 0.26, or

$$0.8 \times 29 \times 0.26 = 6 \text{ FTL}$$

This method is used for all other values.

Step 7. The brightness ratio is found by dividing the common logarithm of the maximum brightness value by the common logarithm of the minimum brightness value, as seen from any viewer location, as the viewer looks at successive points L, C, and R. In the illustration analysis (Fig. 4-6), the brightness ratio as seen by the viewer shown is

$$\frac{\log 26}{\log 6} = \frac{1.415}{0.778} = 1.82$$

This means that point R appears 1.82 times brighter than point L to the eye, even though the nonlogarithmic arithmetic would show $26 \div 6 = 4.33$ times brighter. Brightness ratios not exceeding 3 to 1 are considered satisfactory. The logs may be looked up in a table, read from a slide rule, or called up on a pocket calculator so equipped.

Once the numerical data is obtained, the final results may be displayed in any appropriate way. If a graphic analysis is desired, then a plot of brightness ratio versus view position numbers, for each zone, is appropriate. A horizontal line drawn across the chart at the 3:1 brightness ratio value will quickly show which positions fall outside the good viewing area.

Fig. 4-7 shows how such a graph might look, comparing three competitive screens, at Zone I viewer positions 1, 2, 3 and 4. It should be remembered that the brightness ratio is a logarithmic ratio, as explained previously. The graph is best displayed on graph paper with a vertical log scale, and a horizontal uniform scale. In interpreting the graph, we see that all three screens analyzed in Fig. 4-7 give satisfactory brightness for positions 1 and 2 in Zone I. Screen X is marginal for position 3, and unacceptable for position 4, while screens Y and Z are acceptable. It is obvious that screen Z is the best screen. If a system line coincided with the lower edge of the

aph, where the brightness ratio equals 1.0, it would indi-
e that the screen is perfect, and appears equally bright
m all viewing positions. Similar graphs are now drawn for
 other zones, and the system analysis is complete.

HAT KINDS OF DIFFUSION PATTERNS
) FRONT SCREENS EXHIBIT?

Front screens do not differ nearly as much as do rear
eens, from manufacturer to manufacturer. For example,
 three types of screen material shown in Fig. 4-8 are found
 be very typical. Due to the long throw distance in front
reen arrangements, front screens are usually tested only at
 center point. The longer the throw, the less difference
re will be between the three target points L, C, and R,
ch as were used in the rear screen analysis (Fig. 4-5).

Three conclusions can be drawn immediately from Fig. 4-8:

1. The matte screen, A, has a low figure of merit, but offers fairly uniform brightness over a + and − 50° arc of viewing area.
2. The lenticular screen, B, directs more light to the central audience area, still maintaining good brightness at the sides.
3. The beaded screen, C, is a high-Q, extremely "hot" screen, suitable only for a narrow angle of viewing, like 30° total arc.

In discussing front projection screens, it should be remembered that many such installations are made with the loudspeakers behind the screen, necessitating the use of a perforated screen. Perforating reduces the light reflected to the

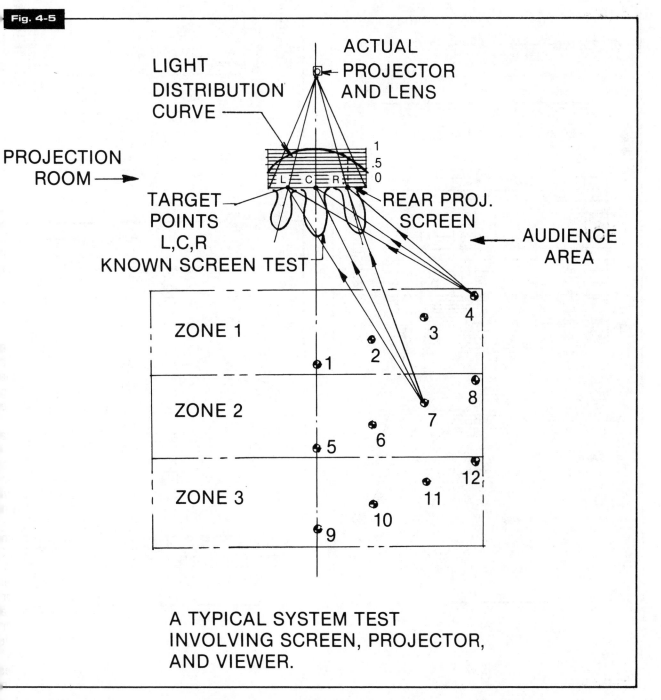

Fig. 4-5

A TYPICAL SYSTEM TEST
INVOLVING SCREEN, PROJECTOR,
AND VIEWER.

Fig. 4-6

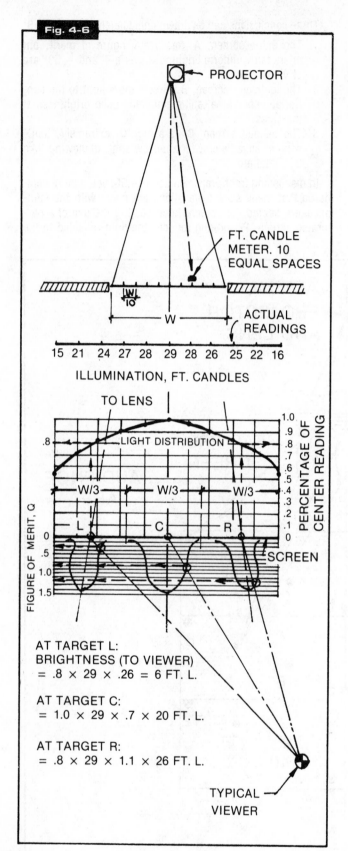

PROJECTOR

FT. CANDLE METER. 10 EQUAL SPACES

$\frac{W}{10}$

W

ACTUAL READINGS

ILLUMINATION, FT. CANDLES

| 15 | 21 | 24 | 27 | 28 | 29 | 28 | 26 | 25 | 22 | 16 |

TO LENS

LIGHT DISTRIBUTION

.8

PERCENTAGE OF CENTER READING

W/3   W/3   W/3

L   C   R

FIGURE OF MERIT, Q

SCREEN

AT TARGET L:
BRIGHTNESS (TO VIEWER)
= .8 × 29 × .26 = 6 FT. L.

AT TARGET C:
= 1.0 × 29 × .7 × 20 FT. L.

AT TARGET R:
= .8 × 29 × 1.1 × 26 FT. L.

TYPICAL VIEWER

Fig. 4-7

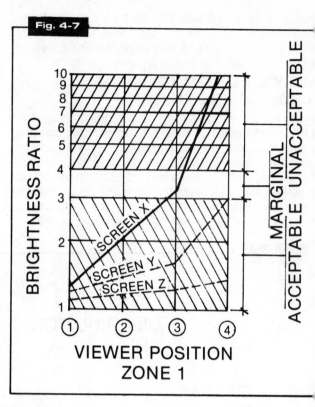

BRIGHTNESS RATIO

SCREEN X
SCREEN Y
SCREEN Z

① ② ③ ④

VIEWER POSITION ZONE 1

MARGINAL
ACCEPTABLE
UNACCEPTABLE

Fig. 4-8

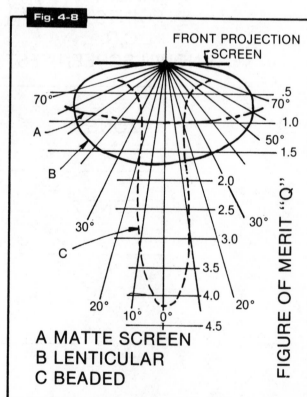

FRONT PROJECTION SCREEN

70°   70°
                .5
A   1.0
50°
B   1.5
               2.0
               2.5
30°   30°
               3.0
C
               3.5
               4.0
20°   20°
10° 0°
               4.5

A MATTE SCREEN
B LENTICULAR
C BEADED

FIGURE OF MERIT "Q"

audience by 10-15%, depending on the size and spacing of the holes. Also, a minimum front row distance of 15 ft is necessary to avoid the distraction of seeing the holes.

WHY DO CURVED SCREENS DIRECT MORE LIGHT TO MORE OF THE AUDIENCE AREA THAN DO FLAT SCREENS?

When a curved screen is used, as illustrated in Fig. 4-9, the rays of light from the projector are reflected in a somewhat radial pattern, converging at a point called the optimum seat. A viewer seated at this location will see the entire screen with

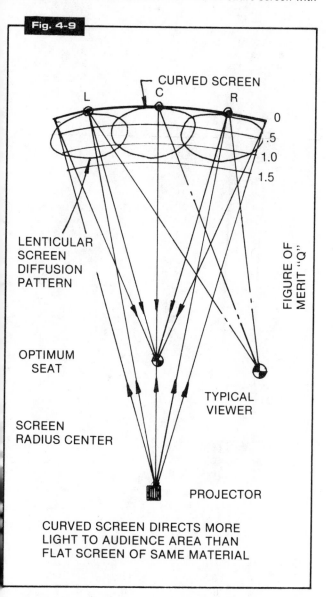

CURVED SCREEN DIRECTS MORE LIGHT TO AUDIENCE AREA THAN FLAT SCREEN OF SAME MATERIAL

uniform brightness, and other viewers nearby will enjoy practically the same brilliance. The typical viewer of the illustration, for example, when observing the three target areas of the screen, L, C, and R, has a sightline that intersects each diffusion curve at a Q of approximately 1.2. The flat screen, unless it is a low-Q matte screen, tends to direct its principal rays in a diverging pattern, not converging at an optimum seat point. Consequently, there is not as great an audience area of sustained brilliance as there is with the curved screen.

WHAT OTHER CHARACTERISTICS AFFECT SCREEN PERFORMANCE?

We have primarily discussed the light diffusion characteristics of projection screens, because this is certainly the most important feature of any screen material. However, every material exhibits other properties which can affect its performance. The most important of these is the ability, or more aptly, the inability, of the screen to reflect ambient light. A low reflectance factor is, therefore, desirable. For example, if a rear screen had a reflectance factor of 0.10, then it would only reflect $\frac{1}{10}$ of the ambient light it received. Thus, if the audience area ambient light were 10 footcandles, the no-image screen brightness would be only 1 footlambert.

A projector illuminating the screen with 30 footlamberts at a given point of observation would, under the above circumstances, produce a ratio of image brightness to stray light brightness of $30 \div 1 = 30$. Ratios of 100:1 are sometimes necessary for good contrast when the projected material contains a long gray scale, or a full range of colors with three dimensional details. Simple graphics with contrasting light and dark areas, black and white as well as color, require as little as a 5:1 ratio. Typical AV media slides, with full range of color, or gray scale, but two-dimensional and, hence, no shadow detail, fall in between and require a 25:1 contrast ratio.

But even the 25:1 ratio is difficult to achieve with high ambient light striking the screen. The ideal aim then, whether front or rear screens are used, is to keep the ambient light from striking the screen, and use a screen of low reflectance. Inasmuch as front screens work by virtue of their reflection capability, it is obviously more difficult to maintain a bright image in the presence of ambient light, than with the rear screen.

Whatever light is not reflected or transmitted is absorbed by the screen. The amount of absorbed light is small, and is usually not a factor to concern us.

Finally, there are a dozen properties of screens, such as fungus resistance, fire resistance, surface characteristics, aging, yellowing, scuff resistance, scintillation, resolution, and so forth, that will not be covered here, as space does not permit.

PROJECTION OPTICS

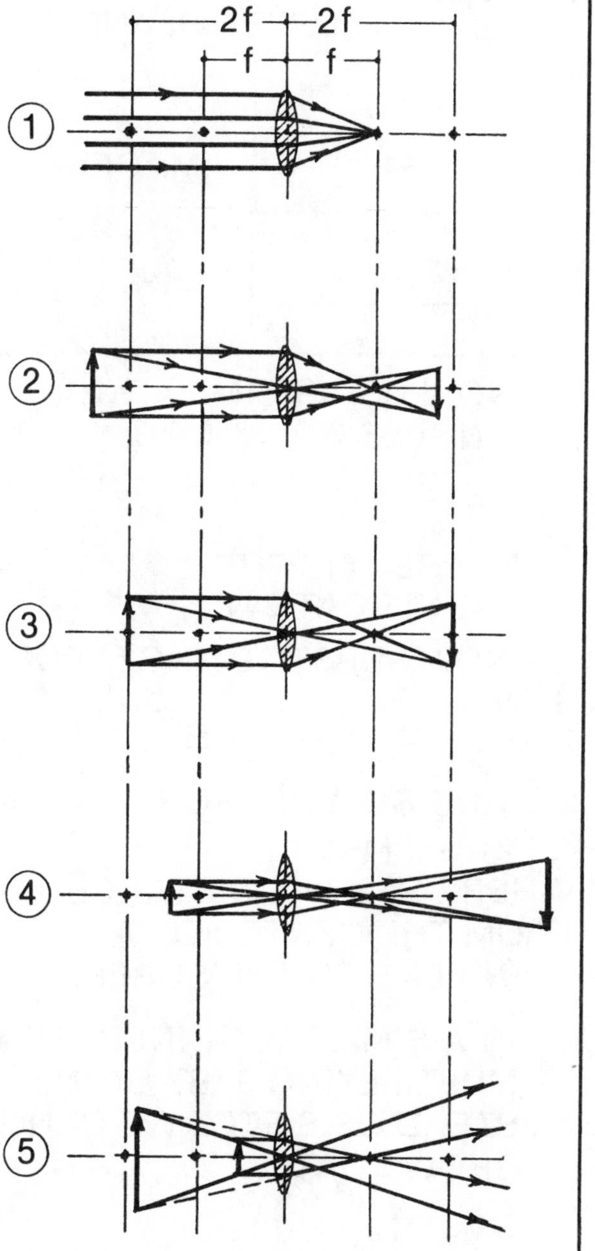

This section concerns itself with information and facts about projection optics which are vital to the design of projection systems. There is evidence of much confusion in published writings regarding such terms as focal length, effective focal length, throw distance, vignetting, curved field, flat field, and many others.

In this section we will illustrate the optical theories that have to do with projection system design. We will analyze published formulas for throw distance, and calculate the effect of focal length on image size.

WHAT ARE THE FIVE WAYS THAT A LENS CAN PRODUCE AN IMAGE OF AN OBJECT?

Fig. 5-1 shows the following:

1. When the object is located an infinite distance from the lens, the emerging light rays are parallel, and they are brought to a focus at the focal point located on the opposite side of the lens. Obviously the lens works the same in either direction.
2. When the object is located anywhere between 2F and infinity, the image is formed on the opposite side of the lens between F and 2F. It is inverted, smaller, and real, which means that it can be seen on a screen placed at the image location.
3. When the object is located at exactly 2F, the image is formed on the opposite side of the lens at 2F. It is inverted, same size, and real.
4. When the object is located between F and 2F, the image is formed beyond 2F on the opposite side of the lens, and is inverted, magnified, and real.
5. When the object is located at a distance less than F, the image appears on the same side of the lens as the object, hence it cannot be projected. Note also that the image is erect, not inverted. Such an image is called virtual, and can be seen directly by the eye. This is the principle of the common magnifying glass.

If the object were to be placed at exactly F, the reverse of Case 1 in Fig. 5-1 would occur, with the outgoing rays parallel. Hence, no image could be formed.

If we now consider that the object is a transparency, or film frame, and the image is a projection screen location, then Case 4 in Fig. 5-1 represents the normal projection situation, wherein a small object (film or slide) is projected magnified onto a projection screen.

The "simple lens" concept shown in Fig. 5-2 is useful in performing certain calculations dealing with object/image relationships in projection optics. It assumes there is a single, thin, convex lens, with its optical plane symmetrically placed at the center of its thickness. Object and image distances S' and S are measured from this plane, and their sum equals the distance from the projector aperture to the screen.

Fig. 5-1

We shall see shortly that when we make use of the more practical "compound lens" formulas, the object and image distances are measured from the front and rear glass surfaces of the lens. The simple lens equation ignores the existence of the nodes of admission and emission in a compound lens, thus introducing an error in the distance from aperture to screen.

While the compound lens equations take this nodal distance into account, it is done so automatically by using the sum of the back focus, BF, front focus, FF, and the glass-to-glass length, L, of the lens assembly, to yield the correct distance between the front and rear focal points of the given lens.

WHAT EQUATIONS CAN WE DEVELOP FOR OBJECT/IMAGE RELATIONSHIP USING A SIMPLE THIN LENS?

Let us study the geometry of the simple lens shown in Fig. 5-2.

Fig. 5-2

THIN LENS GEOMETRY

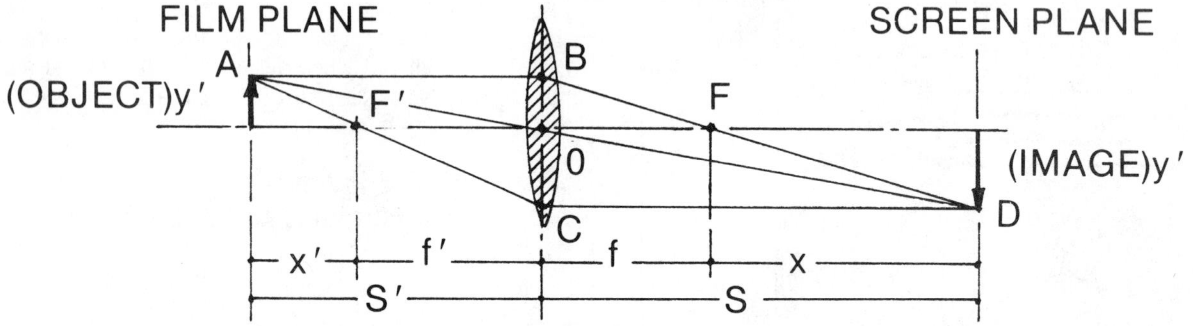

1. $\triangle ABC \sim \triangle F'OC$

2. $\therefore \dfrac{BC}{S'} = \dfrac{OC}{f'}$ OR $\dfrac{y' + y}{S'} = \dfrac{y}{f'}$

3. LIKEWISE, IN $\triangle BCD$ & $\triangle BOF$
   $\dfrac{BC}{S} = \dfrac{BO}{f}$ OR $\dfrac{y' + y}{S} = \dfrac{y'}{f}$

4. ADDING 2 AND 3:
   $\dfrac{y' + y}{S'} + \dfrac{y' + y}{S} = \dfrac{y}{f'} + \dfrac{y'}{f}$

5. BUT $f = f'$

6. $\therefore \dfrac{y' + y}{S'} + \dfrac{y' + y}{S} = \dfrac{y' + y}{f}$

7. FROM WHICH
   $\dfrac{1}{S'} + \dfrac{1}{S} = \dfrac{1}{f}$ THIS IS THE "LENS FORMULA"

8. SOLVING FOR $\dfrac{1}{S'}$:
   $\dfrac{1}{S'} = \dfrac{1}{f} - \dfrac{1}{S}$

9. MULTIPLYING THROUGH BY S: $\dfrac{S}{S'} = \dfrac{S}{f} - \dfrac{S}{S} = \dfrac{S}{f} - 1$

10. NOW USING SIMILAR TRIANGLES AOB & COD:
    $\dfrac{y}{y'} = \dfrac{S}{S'} = M$
    WHERE M IS THE MAGNIFICATION.

11. NOW SUBSTITUTING M FOR $\dfrac{S}{S'}$ IN STEP 9: $M = \dfrac{S}{f} - 1$ OR $\dfrac{S}{f} = M + 1$

12. AND SOLVING FOR S:
    $S = f(M + 1)$
    WHERE S IS THE DISTANCE FROM THE CENTER OF A THIN LENS TO THE SCREEN.

13. IN A SIMILAR MANNER, BY MULTIPLYING 9. BY S' INSTEAD OF S ETC., WE CAN SHOW THAT:
    $S' = f\left(\dfrac{1}{M} + 1\right)$

WHERE S' IS THE DISTANCE FROM THE CENTER OF A THIN LENS TO THE APERTURE.

Note how the rays are formed for the case of the simple lens. Ray A–B is drawn parallel to the optical centerline, so that it will emerge from the lens and pass through the front focal point. This follows from Case 1, shown in Fig. 5-1, which says that any ray emanating, or passing through, the focal point will emerge on the other side of the lens, parallel to the optical axis.

Likewise, ray A–F′–C is drawn through the back focal point F′ so that it will emerge parallel to the optical axis shown in Fig. 5-2. The two rays A–B–F–D and A–F′–C–D meet at point D, forming the image of arrowhead A.

A third ray, A–O–D, drawn through the center O of the lens, also serves to locate point D. In Fig. 5-1, the lens was conveniently drawn large enough to intersect all three rays, but even if it were not that large, the same geometric construction can be drawn by simply extending the vertical centerline B–C as required to intersect rays A–B and A–F′–C. This situation is illustrated in Fig. 5-3.

Equation 9 is called the "lens formula." It is used in elementary lens problems to solve for any one of the three variables when the other two are known. In AV work we have little use for this formula, as we generally prefer to evaluate only the distance on the screen side of the lens which can be done with equation 11.

Equations 11 and 13 appear frequently in published articles on projection optics, but the use of the more practical com-pound lens equations 5 and 7, to be developed, is recommended in AV design calculations.

WHAT EQUATIONS CAN WE DEVELOP FOR OBJECT/IMAGE RELATIONSHIP USING A COMPOUND LENS?

Equations 5, 7, 8, and 9 in Fig. 5-4 are practical equations which take into account the fact that a projection lens contains several elements, resulting in an appreciable lens barrel length. The basic dimensions x and x′ are measured from the front and rear focal points, respectively. The actual throw distance S, and its conjugate S′, are measured from the front and rear glass surfaces of the lens. We do not, therefore, have to know anything about what goes on optically within the lens. This will be made very clear as we solve some problems in projection optics.

HOW DO WE RECONCILE EQUATION 11, SIMPLE LENS ANALYSIS, WITH EQUATION 7, COMPOUND LENS ANALYSIS?

Published literature often shows the "throw distance" for a projection lens as

$$T.D. = f(M + 1)$$

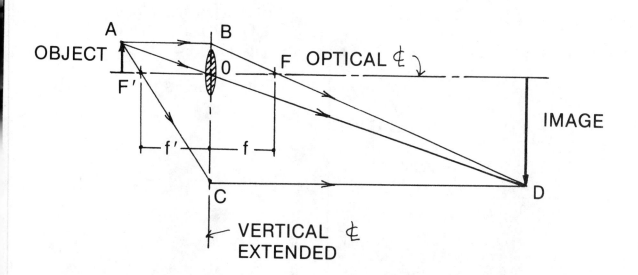

Fig. 5-3

METHOD OF RAY TRACING WHEN RAYS FALL OUTSIDE OF LENS.

where

f = focal length,
M = magnification, which in turn is image height H divided by aperture height h, or image width W divided by aperture width w.

This is equation 11 in the simple lens analysis shown in Fig. 5-2. Likewise, equation 7, compound lens analysis, is often given as the throw distance. Thus

$$T.D. = f(M),$$

where f and M are the same as in the preceding equation.

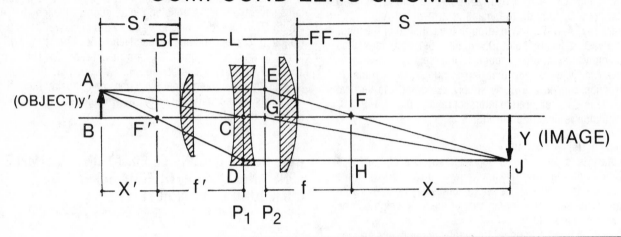

Fig. 5-4

COMPOUND LENS GEOMETRY

1. $\triangle ABF' \sim \triangle F'CD$

2. $\therefore \dfrac{AB}{X'} = \dfrac{CD}{f'}$, OR

3. $\dfrac{y'}{x'} = \dfrac{y}{f}$ (BECASE $f' = f$)

4. LIKEWISE, IN $\triangle EGF$ & $\triangle FHJ$
$$\dfrac{FH}{x} = \dfrac{EG}{f}, \text{ OR}$$

5. $\dfrac{y}{x} = \dfrac{y'}{f}$

6. FROM 3:
$$\dfrac{y}{y'} = \dfrac{f}{x'}$$

7. BUT $\dfrac{y}{y'} = M = $ MAGNIFICATION

8. $\therefore M = \dfrac{f}{x}$, FROM WHICH

9. $x' = f(\dfrac{1}{M})$ WHERE $x' = $ DISTANCE FROM F' TO FILM PLANE.

10. FROM 5: $\dfrac{y}{y'} = \dfrac{x}{y} = M = $ MAGNIFI- CATION

11. FROM WHICH $x = f(M)$ WHERE $x = $ DISTANCE FROM F TO SCREEN.
IT IS NOW OBVIOUS THAT THE THROW DISTANCE IS THE SUM OF FRONT FOCUS FF AND DISTANCE x, OR:

12. $S = x + FF$
THIS IS THE ONLY PRACTICAL WAY TO COMPUTE THROW DISTANCE.

13. LIKEWISE, $s' = x' + BF$

The obvious predicament (actually a misunderstanding of the equations), led one consultant to the conclusion that the second equation was a "short form" of the first, and he warned the AV designer in a published article on projection optics that using the "short form" could result in an appreciable error in throw distance, in cases where the magnification was low. The article further stated that if the magnification was as low as 1 to 1 (M=1), the short form equation would yield an error of 100%! The reasoning behind that statement was obviously that one formula, T.D. = f(M+1), gave a T.D. = f(1+1) = 2f, while the other, T.D. = f(M), gave a T.D. = f(1) = f. Hence, the throw distance from the short form equation produced only half the throw distance given by the "correct" formula! This is an error of 100%. Nothing could be further from the truth! In the first place, T.D. = f(M) is not a shortened form of T.D. = f(M+1), with the +1 merely omitted. As we saw earlier, one formula is for the simple lens; the other is for the compound lens. Both formulas will yield the same throw distance when T.D. is measured from the same point.

Let us take the case of unity magnification, M=1, and calculate the throw distance for both a simple and a compound lens, each with a 5-in focal length. See Fig. 5-5A and B.

It is clear that if we attempted to measure the throw distance of 10 in, as found in Fig. 5-5 for a thin lens, when we are actually using a compound lens, we would have difficulty measuring it. The optical center of a compound lens is not indicated on its barrel, and it is very seldom in the physical center of the barrel length.

It is much more practical to measure from some point that is known, such as from the front vertex of the lens. The front vertex is the center of the frontmost piece of glass in the lens assembly. So we see that the 10-in throw distance given by the thin lens formula is not a practical distance because we do not use a thin lens in a projector, and we do not have a place from which to measure the 10 in when using a compound lens. The 8-in throw distance for the compound lens is a practical distance because we can measure it from the front lens element, to the screen.

Note that for complete accuracy, the front focus distance (FF) must be added to the x = f(M) distance to arrive at the throw distance for a compound lens. But for 2 in and shorter focal length lenses, this distance can be ignored as it is less than 1 in. For longer focal lengths, FF may be taken as approximately ⅔ of the focal length. The lens manufacturer will furnish this information, as well as the back focus BF and lens length L, but it must usually be requested.

Fig. 5-5

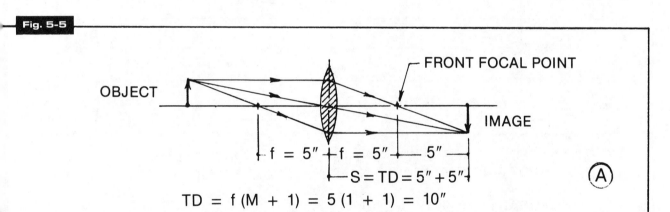

$$TD = f(M + 1) = 5(1 + 1) = 10''$$

P_1 = NODAL PLANE OF ADMISSION

P_2 = NODAL PLANE OF EMISSION

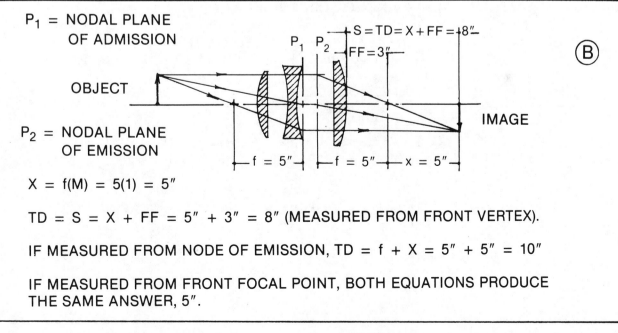

$$X = f(M) = 5(1) = 5''$$

$$TD = S = X + FF = 5'' + 3'' = 8'' \text{ (MEASURED FROM FRONT VERTEX).}$$

IF MEASURED FROM NODE OF EMISSION, TD = f + X = 5" + 5" = 10"

IF MEASURED FROM FRONT FOCAL POINT, BOTH EQUATIONS PRODUCE THE SAME ANSWER, 5".

In example 1, Fig. 5-6, we see that for a compound lens, an accurate distance from the plane of the slide or film to the screen is given by the sum of:

$$x^1 = f\left(\frac{1}{M}\right) = \left(\frac{h}{H}\right)f \quad \text{(all units in inches) This is equation 5, compound lens.}$$

plus BF = back focus distance, inches.

 plus L = length of compound lens, vertex to vertex, inches.

plus FF = front focus distance, inches.

$$\text{plus } x = f(M) = \left(\frac{H}{h}\right)f \quad \text{(all units in inches) This is equation 7, compound lens.}$$

We can see from example 2, shown in Fig. 5-7, that if we are calculating the distance from the slide or film aperture to the screen, both equations will give the same answer, with any slight difference due only to the distance between the two nodes of the compound lens. But if we are calculating throw distance only $(x + FF)$, will be less than $f(M + 1)$ by an amount equal to $(f - FF)$.

Again, because we don't use a simple lens in practice, the compound lens formula $x = f(M)$ should be used, with the FF distance added, to obtain the throw distance measured from the front vertex of the lens to the screen.

WHAT LENS PROBLEMS DO WE HAVE AS A RESULT OF USING REAR SCREEN PROJECTION VERSUS FRONT SCREEN PROJECTION?

Rear screen systems usually require short focal length lenses because of the shallow depth of the projection room. It would be ideal if we could have a projection room deep enough to permit the use of a 3-in or longer focal length lens.

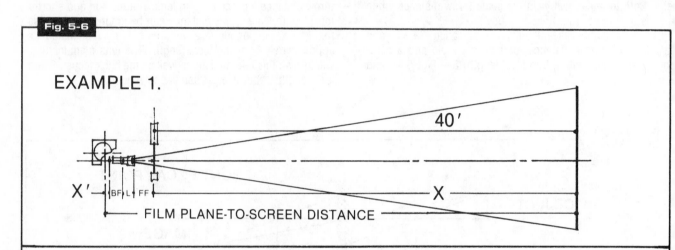

Fig. 5-6

EXAMPLE 1.

40'

X' |BF|L| FF|

X

FILM PLANE-TO-SCREEN DISTANCE

GIVEN: REQ'D IMAGE SIZE 5'-0" HIGH, APPROX. PROJECTION PORT 40' FROM SCREEN. FRONT FOCAL POINT AT PORT.

FIND: FOCAL LENGTH OF ST'D AVAILABLE LENS, EXACT IMAGE SIZE, AND FILM PLANE-TO-SCREEN DISTANCE.

SOLUTION: FROM EQ. 11, COMPOUND LENS GEOMETRY:
X = f(M) = f H/h = f (5/.902) = 40, FROM WHICH f = 7.22"

CHOOSE ST'D 7" FOCAL LENGTH LENS, FOR WHICH
FF = 4.5", BF = 5.1", L = 5" FROM M'F'R'S DATA.

CORRECTED IMAGE SIZE: H/h × 7" = 40', FROM WHICH
H = 40' × .902"/7" = 5.15' = 5'-2", W = 1.48 × 5.15' = 7.63' = 7'-7 1/2"

FIND X' FROM EQ. 9, COMPOUND LENS GEOMETRY.
X' = f (1/M) = f(h/H) = 7" (.902/12" × 5.15) = .102"

FILM PLANE-TO-SCREEN DIST. = X' + BF + L + FF + X
= .102" + 5.1" + 5" + 4.5" + 480" 494.7" = 41'-2 3/4" ANS.

Fig. 5-7

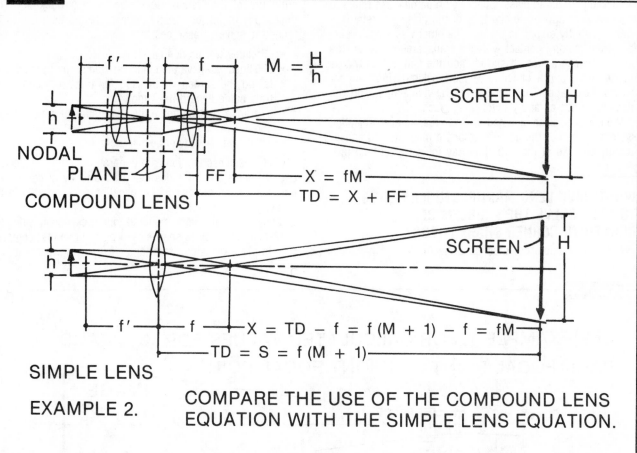

$$M = \frac{H}{h}$$

SCREEN — H

NODAL PLANE

COMPOUND LENS

FF — X = fM

TD = X + FF

SCREEN — H

X = TD − f = f (M + 1) − f = fM

TD = S = f (M + 1)

SIMPLE LENS

EXAMPLE 2. COMPARE THE USE OF THE COMPOUND LENS EQUATION WITH THE SIMPLE LENS EQUATION.

Such is not the case in the majority of installations, and we must settle for a 2-in lens. Lenses of shorter focal length exist, but their use is discouraged where resolution and good light distribution are important.

The shorter the focal length, the more curvature of the lens elements is required, and the more curvature required, the more likelihood of distortion.

The optical systems used in projectors are optimized for the most popular focal length of the projection lens. For 16-mm film projectors, this is a 2-in lens, and for 35-mm slide projectors, it is a 4- or 5-in lens. When lenses of different focal lengths are used, we can expect some loss of illumination. When specifying long or short focal length lenses the designer should check with the manufacturer to see if a compatible set of condensing lenses is available.

The short focal length lenses used in rear projection slide projectors are particularly vulnerable to changes in the flatness of the transparency, due mainly to the heat-warping of the film. This is especially true of slide transparencies when mounted in cardboard or plastic mounts without glass cover plates. The film in such mounts is affected by the heat radiated by the light source. Although much of the infra-red heat energy is filtered out of the light bundle before it passes through the film, enough heat remains to cause the film to buckle, and take the form of a curved surface.

The problem is further aggravated by the more intense heat given off by the high-intensity light sources such as 1200-watt lamps, xenon and ac arc sources.

Sixteen-millimeter motion-picture film, because its aperture area is small, measuring only 0.284 × 0.380 in, is not so readily affected by lamp heat radiation. Also, because the film is in motion, the eye cannot easily assess the amount of distortion present.

It should be remembered that if the plane of the film, for any reason, is displaced forward or backward the slightest amount, the image on the screen will go out of focus. Movement by as little as the thickness of the film will affect image focus. A bulging slide transparency, not mounted between glass covers, can easily deform this much when subjected to heat from the light source.

From what we have already learned, we can quickly put some numbers on this problem, and convince ourselves that such a tiny movement of the film plane will indeed defocus the screen image. Let us use the data established in example 1 (Fig. 5-6). The developed values are summarized in Fig. 5-8.

Obviously, with such a discrepancy in image distance and size, both the center and the edge portions of the slide cannot be in sharp focus at the same time. If we refocus the center, the edges will be out of focus. If we focus on the edges, the center will not be sharp.

This out-of-focus problem with heat-buckled slides becomes worse when short focal length lenses are used. Calculations similar to those just made reveal that the distance x, to the screen, changes less for short focal length lenses, but the image size changes more rapidly, due to the greater spread angle of the light beam.

It is important, particularly in rear screen systems, to hold the transparency flat by mounting it between two thin plates of specially prepared glass. Specially prepared glass refers to the treatment applied to one side of the glass whereby it is ever so slightly etched to prevent the otherwise smooth surface from making contact with the transparency. When the space between a smooth surface and the film is less than a quarter wavelength of light, rainbow colored rings appear across the image. These are called Newton Rings, and the glass is known as anti-Newton Ring glass.

The need for glass mounted slides in rear screen projection systems rules out the use of cardboard mounts, and consequently the use of the 140-slide tray for carousel-type projectors.

WHAT HAVE LENS MANUFACTURERS DONE TO HELP SOLVE THE PROBLEM OF CORNER-TO-CORNER SHARP FOCUS?

A few manufacturers of slide projector lenses have de-signed certain lenses in their line to correct optically for the curvature of the transparency due to heat buckling. This philosophy might just create more problems than it was intended to solve! It assumes that:

- all slides will be nonglass mounted,
- all slides will have the emulsion side of the transparency facing the screen when used for front projection, and thus bulge toward the lamp,
- all transparencies bulge the same amount when heated, and
- a flat, not curved, projection screen is to be used.

While these assumptions are generally true for the amateur use of the majority of slides being used today, it certainly is not true for the professional applications encountered in the industry.

Many slides are glass mounted today, especially since the availability of ready-to-snap-closed plastic mounts, complete with in-place glass covers of anti-Newton ring glass.

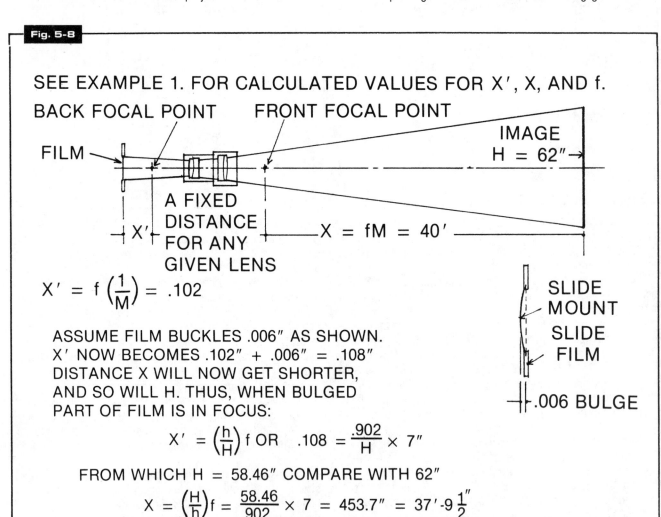

Fig. 5-8

SEE EXAMPLE 1. FOR CALCULATED VALUES FOR X′, X, AND f.

BACK FOCAL POINT — FRONT FOCAL POINT

FILM

IMAGE H = 62″→

A FIXED DISTANCE FOR ANY GIVEN LENS

$X' = f\left(\frac{1}{M}\right) = .102$

$X = fM = 40'$

SLIDE MOUNT
SLIDE FILM

.006 BULGE

ASSUME FILM BUCKLES .006″ AS SHOWN.
X′ NOW BECOMES .102″ + .006″ = .108″
DISTANCE X WILL NOW GET SHORTER,
AND SO WILL H. THUS, WHEN BULGED
PART OF FILM IS IN FOCUS:

$$X' = \left(\frac{h}{H}\right)f \quad \text{OR} \quad .108 = \frac{.902}{H} \times 7''$$

FROM WHICH H = 58.46″ COMPARE WITH 62″

$$X = \left(\frac{H}{h}\right)f = \frac{58.46}{.902} \times 7 = 453.7'' = 37'\text{-}9\frac{1}{2}''$$

COMPARE WITH 40′-0″

FOR THE BULGED CENTER OF THE SLIDE TO BE IN FOCUS, THE SCREEN WOULD HAVE TO MOVE TOWARD THE LENS A DISTANCE OF $40' - 37' - 9\frac{1}{2}'' = 2' - 2\frac{1}{2}''$ CORRESPONDING TO H = 58.46″

While it is true that the emulsion side of the transparency faces the screen when front projecting, the opposite is true for rear projection, unless an image-reversing mirror is used, either within the lens assembly or external to it. But mirrors cause a 4 to 14% light loss, and usually create some loss of resolution. Furthermore, outboard mirrors can often cause ghost images on the screen due to their position in the light beam, are costly, and in some configurations large and bulky. In instances where slides have been copied, or when a re-verse-reading slide is used, the emulsion may be opposite from its normal position. When different emulsions are used, or colored backgrounds, or when the slide is clear except for captions, the amount of heat bulging may vary.

Lastly, some presentations are projected on a flat screen, while others make use of a curved screen. The effect of all the parameters just discussed is especially noticeable when projecting financial data, graphs, charts, and other alpha-numeric data where high resolution is needed, and unfor-tunately, where the viewers are the most critical and demanding.

WHAT IS THE BEST COURSE FOR THE DESIGNER TO TAKE IN DEALING WITH THE ABOVE FOCUS PROBLEM?

The reader should take particular note here that there is a difference between

A. holding the screen fixed and moving the slide axially (forward and backward), and
B. holding the plane of the slide fixed and moving the screen axially.

We learn that these two actions are not reversible. For example, if we axially displace the slide plane a few thou-sandths of an inch, without moving the previously focused lens, the screen has to be moved axially 2 or 3 ft to capture the correct focus. But conversely, if we held the slide fixed in its focused position, and no part of it is displaced due to heat-bulging, we find we can displace the screen axially a couple of feet without defocusing the image appreciably. This situa-tion is similar to "depth of field" in a camera lens. In a pro-jector, however, it is called "depth of focus." It has to do with the eyes' accepting a small "circle of confusion" as the representation of a point. In other words, we accept a slightly blurred image as sharp. This is quite different from accepting part of an image 5 or more inches shorter or longer in size caused by a slight movement of the film plane. In the case of moving the slide slightly, if we additionally moved the screen to bring the image into sharp focus, we would find a tolerance equal to the depth of focus of the lens.

The curvature of a properly designed curved screen is always less than the depth of focus, even when flat field lenses are used.

HOW CAN WE TELL IF A PROJECTION LENS IS OPTICALLY GOOD?

There are several varieties of lens-testing slides which show resolution, geometric distortion, astigmatism, etc., but they are more useful for projector set-up and overall system performance than they are for testing the actual lens quality. In testing a lens mounted in a projector, we must be aware that we are really testing a system, not just a lens. What we see on the screen is the sum total of all the optically related parameters, including the light source and its alignment, the condenser system, the quality of the test slide, the alignment of the lens barrel, the perpendicularity of the optical axis with the center of the screen, the type of screen surface, the ambient illumination, and the visual competence of the viewer.

It is only after optimizing all of the parameters mentioned above that we can pass judgment on the optical performance of the given lens, as evidenced by its ability to show satis-factory resolution, sharpness, and light distribution.

There are two things that we should not expect a projection lens to do:

1. Produce resolution beyond what is normally considered excellent. In other words, if a lens is imaging 14 or 15 equally spaced lines of captions on a 35-mm horizontal-slide format, and the image is in sharp focus from corner to corner when viewed at 8 times the image height, it may be considered extremely satisfactory. If the same lens will not produce a sharp image of a printed magazine page containing some 68 lines of type, don't fault the lens. It is not capable of that kind of resolution, and neither is the human eye.

2. Distribute light uniformly across the screen. The laws of physics prevent this. There has to be fall-off toward the outer edges of the screen because the light is traveling obliquely through the lens when it passes from a corner of the slide, through the lens, and to the opposite corner of the image. The intensity of such oblique rays of light falling on a screen is reduced by a factor of $\cos^4\alpha$, where α is the angular ray displacement measured from optical center. The expression $\cos^4\alpha$ is known as the cosine[4] law, and in itself causes the light distribution curve to follow a parabolic shape.

If we take equally spaced incident light readings across the horizontal centerline of a projected open-slide aperture, and then plot them, we find that the curve falls off on the ends more than a true parabola would indicate. This condition represents a still further light fall-off called "vignetting," and again is something that is not a lens fault. Fig. 5-9 shows graphically an actual light distribution curve of a high quality 2-in focal length lens, and the vignetting condition at the edges of the image. Fig. 5-10 shows how an oblique light ray undergoes vignetting through a short focal length lens.

WHAT IS THE DIFFERENCE BETWEEN FOCAL LENGTH AND EFFECTIVE FOCAL LENGTH?

There is none! These two terms are used so frequently in the literature that one would think that they had different meanings. The focal length of a lens is the distance from the node of emission to the image of a distant object formed by the lens. The most familiar illustration of focal length occurs when we focus the rays of the sun to a point to create a spot hot enough to ignite flammable material.

If we turn the lens around so that the back element now faces the sun, we can produce the same result. A lens thus exhibits the same focal length whether we use it backwards or forwards.

If we had a projection lens whose barrel was engraved 5" f:3.5, it means that the focal length is 5 in and the aperture stop is fixed at 3.5. We can aim this lens at the sun, or at a distant object at least 600 ft away, focus the image on a sheet of white paper, and measure the 5-in focal length from the paper to a plane inside the lens barrel, marking the plane location on the outside of the barrel. This will locate the node of emission. Now if we rotate the lens end for end, as diagrammed in Fig. 5-11, and aim the back end at the object, and repeat the same procedure, we will locate the node of admission.

The two nodal planes thus located may be ¼ in or more apart, or may actually overlap each other. The distance between them is known as the internodal distance. Their significance is that they are the apparent planes from which the light rays seem to emanate when the lens is forming an image of a given object. The magnification between the nodal planes is always 1:1. It is possible therefore to draw geometric object and image rays without actual knowledge of how each lens element refracts the rays. Fig. 5-12 shows nodal planes and ray path for a simple object and image. With a simple thin lens the nodes of admission and emission coincide at the center of the lens thickness. Everything else remains the same.

The term "effective focal length," EFL, came about because each element in the compound lens had its own focal length, but when combined, there is only one focal length, and it was natural to think of this as an effective focal length. The adjective "effective" has been dropped, and the correct terminology is simply "focal length."

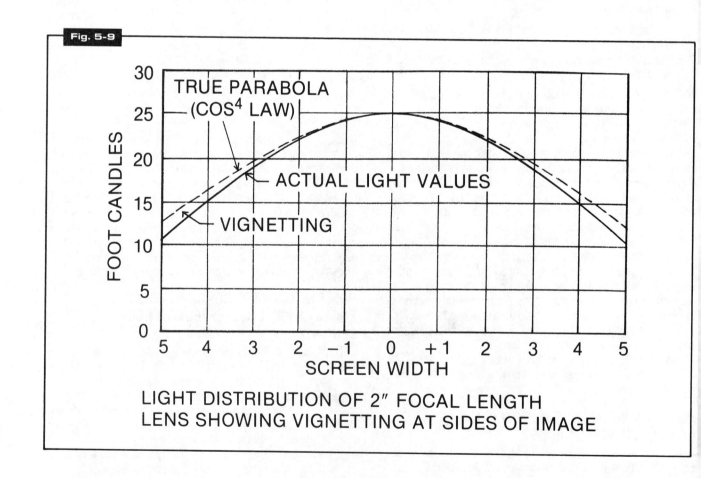

LIGHT DISTRIBUTION OF 2″ FOCAL LENGTH
LENS SHOWING VIGNETTING AT SIDES OF IMAGE

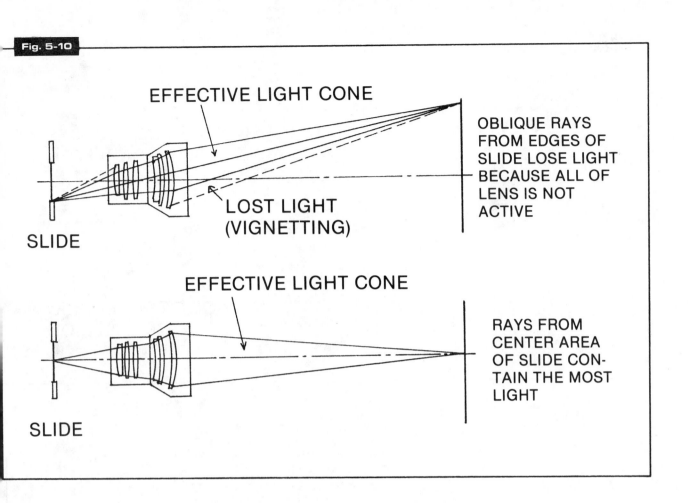

Fig. 5-10

EFFECTIVE LIGHT CONE

OBLIQUE RAYS
FROM EDGES OF
SLIDE LOSE LIGHT
BECAUSE ALL OF
LENS IS NOT
ACTIVE

LOST LIGHT
(VIGNETTING)

SLIDE

EFFECTIVE LIGHT CONE

RAYS FROM
CENTER AREA
OF SLIDE CON-
TAIN THE MOST
LIGHT

SLIDE

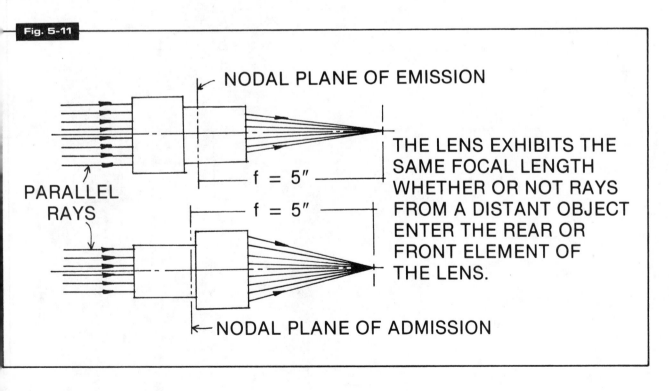

Fig. 5-11

NODAL PLANE OF EMISSION

PARALLEL
RAYS

f = 5″

f = 5″

THE LENS EXHIBITS THE
SAME FOCAL LENGTH
WHETHER OR NOT RAYS
FROM A DISTANT OBJECT
ENTER THE REAR OR
FRONT ELEMENT OF
THE LENS.

NODAL PLANE OF ADMISSION

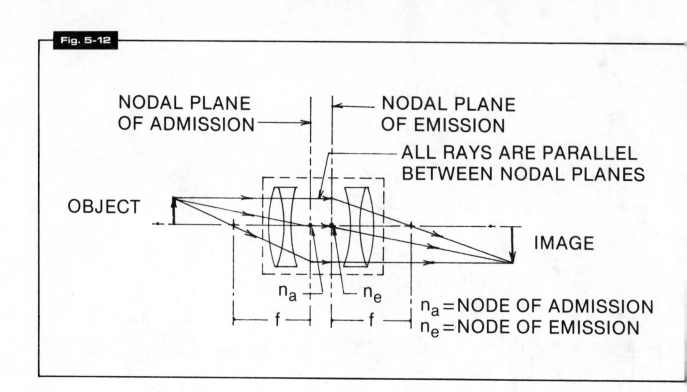

LIGHT OUTPUT OF PROJECTORS

The light output of a projector is one of its most important features—yet it is the one piece of information that is found to be the most obscure. Why is this? Unfortunately there are many reasons why manufacturers are reluctant to advertise the guaranteed light output of a projector. Here are a few of them, along with some relevant comments:

1. Light output of a projector is measured in lumens, and the lumen output can become a "numbers" game just as the horsepower of automobile engines was a few years ago. Today, automobile makers stress economy, or miles per gallon, and projector manufacturers talk about "brilliance" and corner-to-corner screen illumination. We see nebulous claims such as "twice as much light," "unmatched performance," "brighter, sharper images," and so on. But these are all relative terms, and provide little help to the designer who is trying to design an audio-visual projection system.

2. If a manufacturer advertises "x" number of lumens when his projector is projecting a 35-mm double frame slide, then a competitor could claim almost twice as many lumens with the same lens and lamp, by the simple expedient of not mentioning the type of slide, and running his test using a superslide. The superslide, while fitting the same 2- × 2-in slide mount, has an aperture measuring $1\frac{1}{2} \times 1\frac{1}{2}$ in while the double frame slide aperture is 1.34×0.902 in. Inasmuch as the projector is designed to use both formats, the light is available, and provided that the lens entrance pupil is completely filled by the light beam, the larger format will produce almost twice as much light. It is therefore common practice for manufacturers to advertise, say, 4000 lumens as the output of a certain xenon slide projector, without warning the designer that such a rating is based on the use of a superslide! It may come as a surprise to find out after installation that only 1900 lumens are reaching the screen through a 35-mm slide format.

3. Again, two competitive projectors, identical except for the f/stop of the projection lens, may show an appreciable difference in light output, depending on how completely the condenser light beam fills the lens aperture. For example, if we compare a 2-in f/2.8 lens with a 2-in lens having an aperture of f/2.5, the latter will transmit $2.8^2 \div 2.5^2$, or 1.25 times the light transmitted by the f/2.8 lens. Because a manufacturer has no control over what lens the user may decide to install, he does not want to be responsible for claiming light outputs that are unobtainable. A case in point is the user who buys a 16-mm projector advertised to emit 1200 lumens. It is furnished new with a 2-in focal length lens. The buyer then proceeds to purchase a $\frac{5}{8}$-in focal length lens so that he can use the machine in a rear screen projection system. To his dismay he finds

that the projector is producing only 480 lumens—not enough to illuminate his 60 sq ft screen to more than 9 lumens per square foot! He had counted on 20 (1200 ÷ 60). Note that short focal length lenses lose a lot of light because their entrance pupils are small, and they are located close to the projector aperture. Much of the light from the condenser never gets through the lens.

4. Some projection manufacturers advertise brightness ratings for their projectors, but if we read the fine print, we see that light meter readings were taken using a screen with a gain of 2. We should not really be concerned with brightness when we are talking about screen illumination. Brightness is a matter of screen performance, and illumination is a matter of projector-light output. The screen does not enter into illumination readings. Here is a case where the manufacturer is trying to capitalize on the gain of the screen to imply high light output from his projector.

The best way to obtain reliable light output data for various projectors, light sources, and lenses is to run a test using the exact equipment specified, and accumulate such data in a design file. The results of such tests will seldom agree with published data, but that is the reason we recommend such a procedure.

This section will explain the terminology and methods used to test projectors for light output, using readily available equipment and simplified procedures. A sample test data sheet will be presented.

WHAT ARE THE BASIC UNITS OF MEASUREMENT USED IN PROJECTOR LIGHT OUTPUT TESTS?

There are three basic units used in light measurement: the *lumen,* the *footcandle,* and the *footlambert.* We mentioned these units briefly in Sections 4 and 5, but we will now examine them in more detail. Although the footlambert is not used for measuring incident light (illumination), it is included here for completeness.

The Lumen

The lumen is a measure of the quantity of light emitted from a light source. The original standard light source was, logically enough, a wax candle made to precise specifications, and called a "standard candle." The light spread by the flame of this candle uniformly over a 1-sq ft area of a 1-ft radius hollow sphere surrounding the candle, with its flame at the center, was called a lumen. Thus the standard candle actually produced 12.57 lumens of light, inasmuch as the interior surface of a hollow sphere is $4\pi r^2$, or $4\pi \times 1^2 = 12.57$ sq ft. The term standard candle has now been dropped, and the *candela,* pronounced can-del'a, has replaced it as the standard unit of luminous intensity.

The Footcandle

To avoid confusion between lumens and lumens per square foot, the name footcandle has been given to lumens per square foot. As an example of the use of these two units, we may say that a projection screen measuring 8 ft wide × 5.4 ft high, receiving an illumination of 800 lumens from a projector, has a light density averaging 800 ÷ (8 × 5.4) = 18.5 footcandles (lumens per square foot).

The Footlambert

Although the screen receives 18.5 footcandles of illumination from the projector, this is not the intensity the viewer sees. Some of the light is absorbed by the screen, and the rest is either transmitted or reflected, depending on whether the screen is a rear, or front projection screen. Rear screens not only transmit, but also reflect a portion of the light, so what remains to be viewed is called the *screen brightness,* and is measured in footlamberts.

The footlambert, like the footcandle, is actually a "lumen-per-square-foot" type of unit. Its main distinction is that it measures the brightness of a source surface, rather than the light incident on a given surface.

Fig. 6-1 shows graphically the basic light-measuring terms.

WHAT "LAWS" ARE INVOLVED WITH LIGHT MEASUREMENT? EXPLAIN THEIR USE.

Three so-called laws are of prime importance in light measuring work. They are:

1. The inverse square law,
2. Lambert's cosine law, and
3. The cosine[4] law.

The Inverse Square Law

This law is intuitive. It states simply that the further away we place a light source, the less the illumination that is received. Mathematically it states that the illumination from a light source varies inversely as the square of its distance from the measuring point.

There are many who confuse this situation with putting a projector at the rear of an auditorium, versus placing it nearer the screen. Actually the inverse square law does not apply here because we are talking about the same size image on the screen—we are not letting the image grow larger as the projector is moved back. Hence the image will not grow dimmer, as we will be funneling the same amount of light through a longer focal length lens at the far position to maintain the same image size.

Note that in Fig. 6-2, if we don't change the focal length of the lens, and simply double the distance, the image will be four times larger, and one-fourth as bright. If we move the image to 60 ft from the projector, the image will be nine times larger and one-ninth as bright.

Lambert's Cosine Law

Lambert's cosine law is used to allow for the obliquity of an angular ray when striking a flat surface which is perpendicular to the optical axis of the light beam as shown in Fig. 6-3. Light meters are cosine-corrected so that they allow automatically for the lesser intensity of illumination on a flat

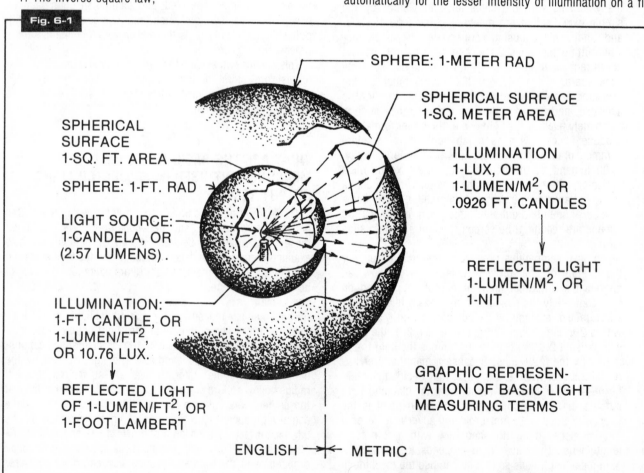

SPHERE: 1-METER RAD

SPHERICAL SURFACE 1-SQ. METER AREA

ILLUMINATION 1-LUX, OR 1-LUMEN/M² , OR .0926 FT. CANDLES

REFLECTED LIGHT 1-LUMEN/M² , OR 1-NIT

SPHERICAL SURFACE 1-SQ. FT. AREA

SPHERE: 1-FT. RAD

LIGHT SOURCE: 1-CANDELA, OR (2.57 LUMENS).

ILLUMINATION: 1-FT. CANDLE, OR 1-LUMEN/FT² , OR 10.76 LUX.

REFLECTED LIGHT OF 1-LUMEN/FT² , OR 1-FOOT LAMBERT

GRAPHIC REPRESENTATION OF BASIC LIGHT MEASURING TERMS

ENGLISH ← → METRIC

Fig. 6-1

surface when the light ray is at an angle. Consequently the light meter is held flat against the screen when taking incident light readings.

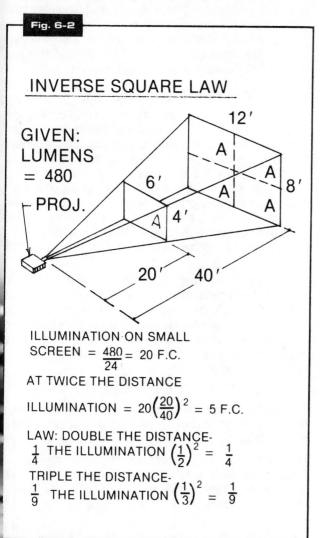

Fig. 6-2

INVERSE SQUARE LAW

GIVEN:
LUMENS
= 480
PROJ.

ILLUMINATION ON SMALL
SCREEN = $\frac{480}{24}$ = 20 F.C.

AT TWICE THE DISTANCE

ILLUMINATION = $20\left(\frac{20}{40}\right)^2$ = 5 F.C.

LAW: DOUBLE THE DISTANCE-
$\frac{1}{4}$ THE ILLUMINATION $\left(\frac{1}{2}\right)^2 = \frac{1}{4}$

TRIPLE THE DISTANCE-
$\frac{1}{9}$ THE ILLUMINATION $\left(\frac{1}{3}\right)^2 = \frac{1}{9}$

The Cosine⁴ Law

This law takes into account the luminance (brightness) of the source, the resulting illumination on a normal (perpendicular) surface at a given distance, and the illumination at some point making an angle α with the optical center. Referring to Fig. 6-4, the law states that the footcandle reading at point "B" is equal to the footcandle reading at point "A" multiplied by the 4th power of the cosine of angle α.

The cosine⁴ law explains why, when a cone of light passes through the center of a lens, the light intensity falls off toward the corners of the screen. This light fall-off is not a lens fault, but merely a law of physics at work.

ARE VIGNETTING AND COSINE⁴ LAW LIGHT LOSS THE SAME THING?

No. We saw in Section 5, Projection Optics, that a plot of the light distribution along the horizontal centerline of a screen showed how the shape of the vignetted distribution curve (parabolic for practical purposes), differed from the cosine⁴ law curve. It is emphasized again that neither cos⁴

light fall-off nor vignetting light fall-off is lens faults. Any advertising that claims "even" light distribution across the screen is obviously not to be believed.

WITH OUR KNOWLEDGE OF LIGHT DISTRIBUTION THROUGH A LENS, IS IT POSSIBLE TO "SYNTHESIZE" A LIGHT DISTRIBUTION CURVE?

Yes, and such an exercise is an extremely valuable undertaking. It is perhaps the only way the reader will gain a full understanding of the way the lens distributes the light over the screen area. The method for doing this is carefully described here in a step-by-step analysis, including a full graphical treatment. A 35-mm slide projector is assumed, capable of projecting 35-mm double-frame slides and super-slides.

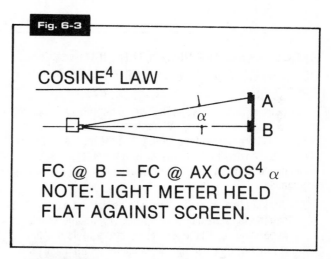

Fig. 6-3

COSINE⁴ LAW

FC @ B = FC @ AX COS⁴ α
NOTE: LIGHT METER HELD
FLAT AGAINST SCREEN.

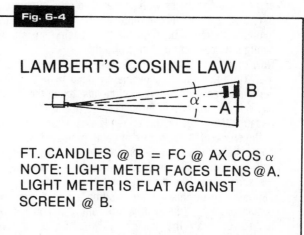

Fig. 6-4

LAMBERT'S COSINE LAW

FT. CANDLES @ B = FC @ AX COS α
NOTE: LIGHT METER FACES LENS @ A.
LIGHT METER IS FLAT AGAINST
SCREEN @ B.

STEP 1. THE CONDENSER LIGHT BEAM MUST COVER THE DIAGONAL OF A SUPERSLIDE.

Fig. 6-5 shows the relationship between the condenser light beam and the two slide formats, 35-mm df and super-slide. The dashed circle represents the minimum beam size at the slide projector aperture that will cover the superslide format. Sometimes the beam is shaped as shown by the dashed line, and the condenser elements are, of course, cut the same way to allow a more compact lamphouse. This truncating of the condenser lens does not affect the working portion of the light beam, nor does it affect any of our calculations.

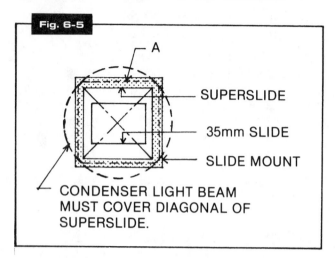

Fig. 6-5

A

SUPERSLIDE

35mm SLIDE

SLIDE MOUNT

CONDENSER LIGHT BEAM MUST COVER DIAGONAL OF SUPERSLIDE.

STEP 2. DRAW THE PROJECTOR LIGHT BEAM IN RELATION TO THE PROJECTED APERTURE

A feature of this analysis is that we do not need a projector, a screen, or a light meter! All we need is the basic cosine[4] law, a little knowledge of vignetting, and the means of making a graphical analysis (compass, scale, protractor, pencil, and paper). The drawing in Fig. 6-6 is developed in the following sequence:

A. Working to a scale of twice full size, lay out a 3- × 3-in square to represent the 1½- × 1½-in aperture of a superslide. Locate its center by crossing its diagonals.
B. Divide half of one of the diagonals into eight equal parts, and draw eight circles with consecutive radii numbered 0-1, 0-2, 0-3, etc. These represent rings of uniform illumination. Remember that the light beam consists of contiguous concentric circles of light.
C. Now draw a rectangle representing the 35-mm df format. Note that this is conveniently done by letting the vertical tangents to the No. 5 circle just touch the No. 6 circle at the upper and lower corners of the format.
D. Because the image on the screen is exactly proportional to the shape of the slide format being projected, we can now consider the aperture formats we have drawn to be actual screen projections. We can draw a plan view of the screen, and the projected light beam as shown in Fig. 6-6. To do this we will assume a rear projection system using a 2-in focal length lens. Section 5 furnishes us with the necessary equation for throw distance.

We will assume that the screen image represents a scale of ¼ in equals 1 ft-0 in, and draw the lens in its calculated position.

$$\text{Throw distance} = \frac{H}{0.902''} \times 2'' \text{ FL}$$
$$= \frac{W}{1.34''} \times 2'' = 15.67'$$
$$= 15'\text{-}8'' + 2'' \text{ to apex of beam}$$

Note that the width of the double frame image on the screen 10 ft-6 in wide to the scale just given. The eight divisions the left half of the light beam have been projected vertical onto the line representing the screen, marked 0 to 8, and connected to the lens apex. These lines form the oblique angle of incidence, α, that each ray makes with the normal center line 0-S. The angles are measured with the protractor, or calculated, and indicated on the drawing (Fig. 6-6).

STEP 3. APPLYING THE COSINE[4] LAW

What we have so far is a projector lens, a light beam, and screen, with a 35-mm df slide format projected on it. Th width of the image spans 10 equal divisions, and if extende to 16 divisions it would cover a light beam capable of span ning the diagonals of a superslide.

If we took a footcandle meter and traversed the screen (o the side facing the projector) along the horizontal centerline o the screen, recording a reading at every division mark, w would find the highest reading at the center position "0, with gradually diminished readings at each successive divi sion.

But this is exactly what the cosine[4] law tells us. It says tha if we know the center reading, and the angle, α, that an point removed from the center makes with the centerline 0-S then we can calculate the illumination at that point by the us of the law. For example, if the illumination at point "0" wer 100 footcandles, referring to Fig. 6-6, then the illumination a point "2" would be:

$$I_2 = 100 (\cos^4 7.6°) = 96.5 \text{ footcandles}$$

The accompanying table in Fig. 6-7 gives the result of thi calculation for the points 1–8, assuming that the center read ing is 100 footcandles.

The reason for using a center value of 100 is that when a readings are found, they will represent percentages of th center reading. Note that vignetting effects have been esti mated where applicable, points 5 through 8.

Example:
A center-screen light-meter reading shows 28 footcandles What reading might we expect at a point having an angle o incidence of 15°?

Solution:
The table in Fig. 6-7 shows that at 15° (point 4), 87.1% o the center screen illumination is being transmitted. Therefore,

$$I_{15} = 28 \times 0.871 = 24.4 \text{ footcandles}$$

STEP 4. PLOTTING THE LIGHT DISTRIBUTION CURVE

It is convenient now to reproduce the 35-mm and super slide formats with the circular divisions made in Step 2, and to construct directly above them a graph showing the il lumination in footcandles for each division point. The foot candle values are those tabulated in Step 3, and are plotted as shown in Fig. 6-8 as percentages of the center value. Vignet ting may be expected to attenuate the light as shown by the dashed line. The shape of the vignetted curve is interesting, and if we analyze it, it will be found to be surprisingly close to that of a parabola, and for practical purposes may be as sumed as such.

Fig. 6-6

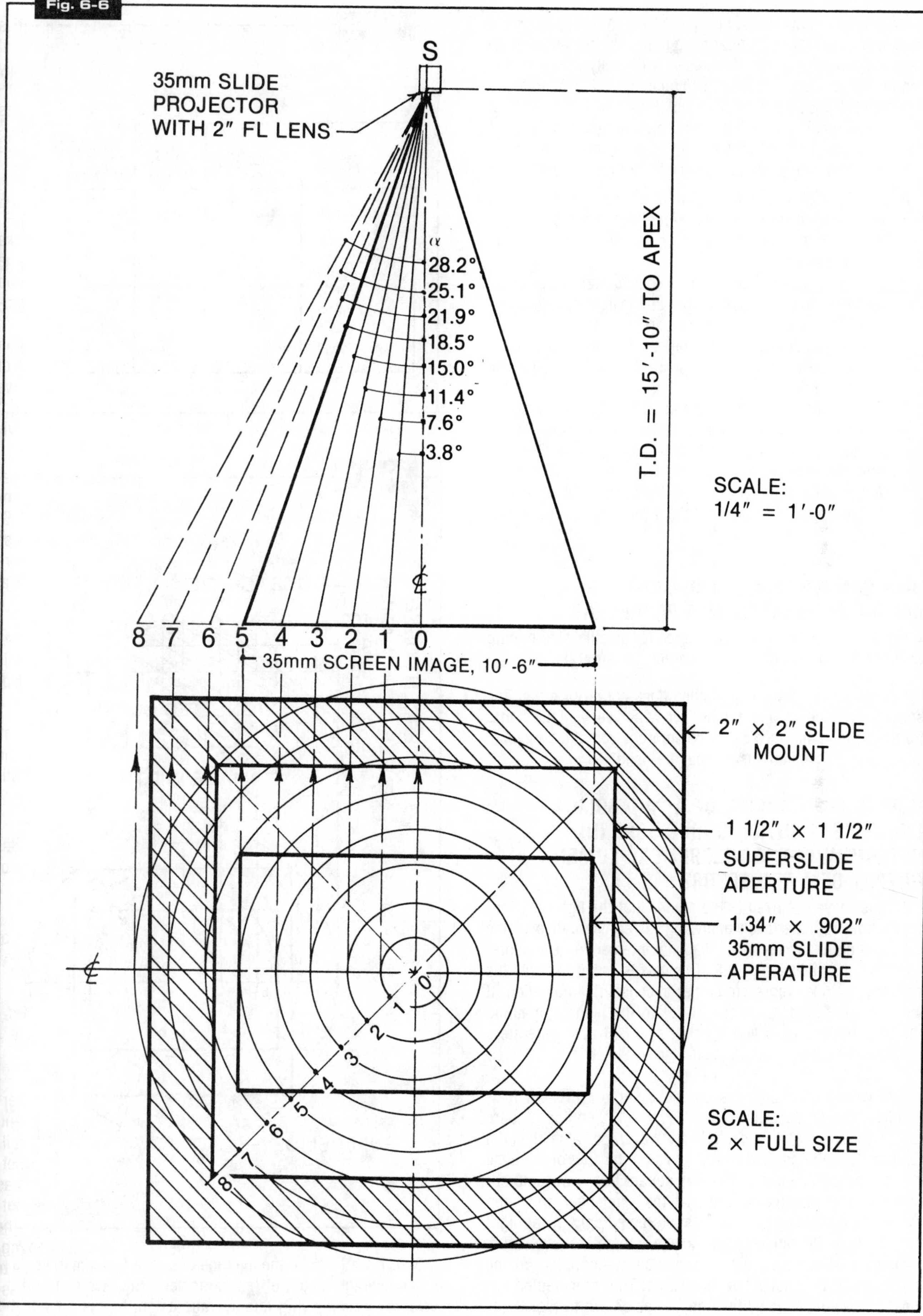

S

35mm SLIDE
PROJECTOR
WITH 2" FL LENS

α
28.2°
25.1°
21.9°
18.5°
15.0°
11.4°
7.6°
3.8°

T.D. = 15'-10" TO APEX

SCALE:
1/4" = 1'-0"

8 7 6 5 4 3 2 1 0

35mm SCREEN IMAGE, 10'-6"

2" × 2" SLIDE
MOUNT

1 1/2" × 1 1/2"
SUPERSLIDE
APERTURE

1.34" × .902"
35mm SLIDE
APERATURE

0
1
2
3
4
5
6
7
8

SCALE:
2 × FULL SIZE

We can see, graphically, how point "P," near the corner of the superslide format is undergoing both cos⁴ law and vignetting attenuation, being illuminated with only 72% of the center illumination. Similarly the illumination all around circle 5, which includes point "P₁," in the 35-mm format, is 80% of the center reading. Consequently, the illumination at point "P₁" has fallen off 20% compared to the center reading.

Some conclusions we are able to draw thus far are:

A. Fall-off of light from the center of an image is dependent upon the cosine⁴ law, and vignetting. Both conditions are normal, and should not be thought of as faulty optical design.

B. Fall-off in the corners of a 35-mm format is obviously less than that for a superslide, as the light covering this format falls within the upper portion of the distribution curve. Accordingly, light fall-off specifications for a superslide format must be less stringent than for the 35-mm slide.

C. If we want a more favorable fall-off of light from center to corner, then we must increase the diameter of the condenser light beam to something greater than the superslide diagonal.

D. Any specification for corner fall-off cannot be arbitrary, but must take into account the laws of optics that are at work.

HOW CAN WE TELL IF THE LIGHT DISTRIBUTION CURVE IS A PARABOLA?

Fig. 6-9 describes a simple method for quickly determining whether the given light distribution curve is parabolic.

Observe that the preceding method does not require that any specific scale be used for the horizontal and vertical divisions. Further, the vertical scale may be different from the horizontal scale. Any convenient scales may be used.

HOW IS THE CONCEPT OF A PARABOLIC LIGHT DISTRIBUTION CURVE USED IN DETERMINING THE THEORETICAL LUMEN OUTPUT OF A PROJECTOR?

It must have occurred to the reader that the light distribution curve we have been analyzing is representative of the curve taken along any other diagonal intersecting the center "0," such as B-B in Fig. 6-8. This makes sense because the concentric circles represent circles of constant illumination. It would be possible, therefore, to rotate the plane containing the distribution curve A-H-A about axis 0-H and generate a paraboloid of revolution as illustrated in Fig. 6-10.

The values plotted on the parabolic curve represent footcandles, and by definition a footcandle is 1 lumen per square foot. It follows that if we could find the average value of all the footcandle readings, and multiply this by the area of the base circle (dashed line in Fig. 6-10) in square feet, we would obtain the total lumens of light that the condenser and lamp system is producing. But we are interested only in the light output that illuminates the image, therefore we must "truncate" the paraboloid of revolution by chopping off the sections that lie outside the image area. The truncated paraboloid of revolution would then appear as shown in Fig. 6-11.

POINT	$\alpha°$	100 COS⁴ α FC	MULTI-PLIER	CORRECTED FOR VIGNETTING
0	0°	100.0	1.0	100.0
1	3.8	99.1	1.0	99.1
2	7.6	96.5	1.0	96.5
3	11.4	92.4	1.0	92.4
4	15.0	87.1	1.0	87.1
5	18.5	80.8	.99	79.9
6	21.9	74.1	.97	71.9
7	25.1	67.2	.93	62.5
8	28.2	60.4	.83	50.1

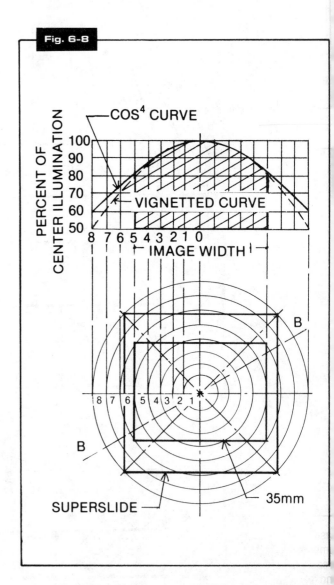

Fig. 6-8

Now if we think of the average of all the footcandle vectors as an average height of the paraboloid, then we can find its volume by multiplying this average height by the area of the

52

Fig. 6-9

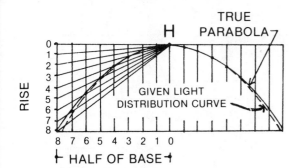

RISE

TRUE PARABOLA

H

GIVEN LIGHT DISTRIBUTION CURVE

8 7 6 5 4 3 2 1 0

⊢ HALF OF BASE ⊣

a. DIVIDE THE RISE INTO SAME NUMBER OF PARTS AS 1/2 BASE.

b. DRAW 8-H, 7-H, 6-H, ETC.

c. DRAW TRUE PARABOLA AT INTERSECTION OF 0-0,1-1, ETC.

d. NOTE SIMILARITY BETWEEN PARABOLA AND ACTUAL LIGHT DISTRIBUTION CURVE.

itself into finding the volume of a truncated paraboloid when the center and edge heights are known, along with the area of the base (screen). This is now a purely mathematical problem, easily done by the methods of calculus. When the calculation is carried out, we find that the average height of the paraboloid of revolution is

$$H_{ave} = C - \frac{C-Z}{3}, \text{ or } H_{ave} = \frac{2C + Z}{3}$$

Either equation will yield the same result. In the preceding equations:

C = center of screen image area illumination, footcandles. Open gate, no film.
Z = corner illumination, footcandles.
A = area of image aperture on screen, W × H sq ft.
W = width of screen image, ft.
H = height of screen image, ft.
V = volume, or total lumens.

Total volume, or lumens, is found by multiplying:

$$\text{Lumens} = WH \left[\frac{2C + Z}{3} \right]$$

This equation tells us that if we have a properly adjusted condenser and light source system in a 35-mm slide projector, we can find the total lumen output by simply taking one corner reading, and the center reading, then substituting these readings in the preceding equation.

Fig. 6-10

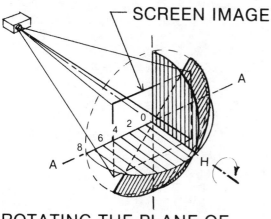

SCREEN IMAGE

A

0 2 4 6 8

A

H

ROTATING THE PLANE OF THE LIGHT DISTRIBUTION CURVE A-H-A ABOUT 0-H PRODUCES A PARABOLOID OF REVOLUTION

Fig. 6-11

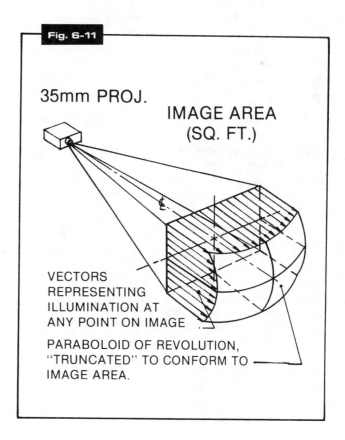

35mm PROJ.

IMAGE AREA (SQ. FT.)

VECTORS REPRESENTING ILLUMINATION AT ANY POINT ON IMAGE

PARABOLOID OF REVOLUTION, "TRUNCATED" TO CONFORM TO IMAGE AREA.

base. The volume will then represent the total light output in lumens, because lumens equal average footcandles times area of screen. We might start logically with a given center and a given corner reading, then find the average reading (height) for the entire solid in terms of these two readings. Note that in Fig. 6-11, a corner reading and a center reading are necessary to define a parabola. The problem now resolves

WHAT IS THE STANDARD METHOD OF MEASURING LIGHT OUTPUT OF PROJECTORS?

The American National Standards Institute, ANSI, formerly known as American Standards Association, ASA, has published certain standards sponsored by the Society of Motion Picture and Television Engineers (SMPTE), describing the preferred method for measuring average screen illumination. The method is illustrated in Fig. 6-12.

In either front or rear projection, the light meter is held flat against the screen, on the projector side of the screen. A cosine-corrected light meter should be used, calibrated in footcandles. Readings are affected by any ambient light present, and if any is present, its value should be noted and subtracted from the incident light reading from the projector. Ambient light is read from the same meter position used for the incident light reading for the projector, but with the projector turned off.

Projector light output tests are extremely valuable for system design data, but in order to be of maximum usefulness, complete data should be recorded. Such variables as lamp voltage, type of lamp and age, condensing lens arrangement, heat filters, projection lens type, f/stop, etc., are all important for an accurate analysis. Once a projector set-up is made, several tests should be run, if possible, to examine the variations caused by changing lenses, lamps, heat filters, and so on. A test made for a 4-in lens is not valid for a 2-in lens.

Prior to taking light readings, a low-density transparency is placed in the gate of the projector, to determine the correct masked edges of the image when the slide is in focus. If a 16-mm projector or other projector using film is being tested, the projected aperture is measured when the image is in sharp focus. The edges of an open aperture, when brought into focus on the screen, do not produce the same size area as that resulting from focusing on the transparency or film. This is true because the aperture plate is not in the same plane as the film. The focus is not changed during the test.

The focused area is divided as explained in the ANSI standard, the slide or film is withdrawn, and light readings taken. In running tests for a rear screen system, where short focal length lenses are required, it is advisable to use a glass-mount slide holder, under the assumption that critical slide focusing will be required, calling for glass-mounted slides. This is especially important when high-intensity light sources are contemplated.

Fig. 6-13 shows a suggested test sheet. The ANSI standard test does not contemplate making a light distribution curve, so provision for recording the necessary data has been made at the bottom of the sheet. Here will be found a horizontal line divided into 10 equal parts. This line represents the horizontal centerline of the screen. A meter reading is taken at each division line and recorded across the plot. These readings are not only required to investigate light distribution by the lens, but are also needed when analyzing screen performance.

If much light output testing is to be done, it saves considerable time to make an opaque slide containing nine properly spaced holes for light-meter location, and four edge marker holes to identify the correct image size. Simply project this slide on a suitable wall or screen, and take readings in the center of each of the nine "holes." Any flat, light-colored surface will suffice for a light output test; the screen is not being tested here. A sample Test Sheet is shown in Fig. 6-13 with all necessary dimensions for making both the superslide and 35-mm df test slide.

Fig. 6-12

ANSI METHOD FOR MEASURING LIGHT OUTPUT OF PROJECTORS

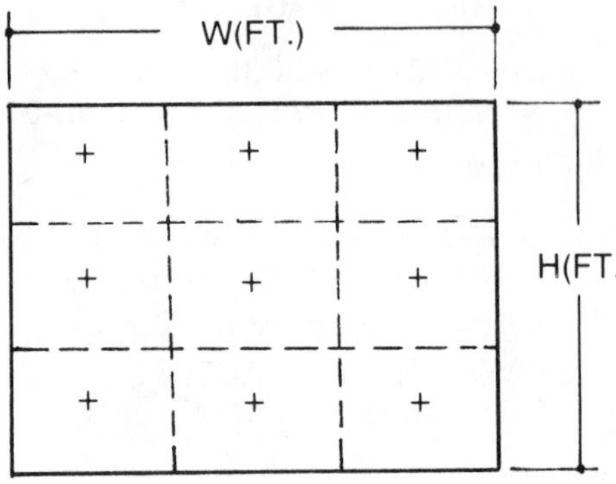

a. DIVIDE SCREEN INTO 9 PARTS.
b. READ FOOTCANDLES AT CENTER OF EACH DIVISION.
c. ADD ALL READINGS AND DIVIDE BY 9 TO FIND AVERAGE.
d. LUMENS = W × H × AVE. F.C.

NOTE: IF W × 43 3/4", H WILL BE 29 5/8". AREA WILL BE 9 FT.^{2} THE 9 IN STEP C. WILL CANCEL OUT, AND THE SUM OF THE 9 READINGS WILL BE THE LUMEN OUTPUT.

THE AUTHOR PREFERS A LARGER SCREEN FOR MORE ACCURACY.

WHAT IS THE ACCURACY OF THE STANDARD ANSI 9-POINT METHOD FOR DETERMINING THE LUMEN OUTPUT OF A PROJECTOR?

To answer this question we will resort to our theoretical analysis (Fig. 6-8) wherein we made use of the light-distribution curve plotted from the cos^4 law with vignetting. We went on to calculate the total lumens mathematically. If we return to that analysis, we will be able to superimpose the ANSI 9-point grid onto our assumed image, and finally compare the results with the mathematical calculation. The method is described here, and Fig. 6-14 shows the graphical layout required.

Fig. 6-13

SAMPLE TEST DATA SHEET FOR RECORDING LIGHT OUTPUT OF PROJECTION DEVICES

PROJECTOR LIGHT OUTPUT TEST

DATE: _____   TESTED BY: _____

PLACE: _____   METER: _____

PROJECTOR:
- ☐ 35mm SLIDE
- ☐ 16mm FILM
- ☐ 35mm FILM
- ☐ OVERHEAD
- ☐ FILM STRIP
- ☐ OTHER

M'F'G   MODEL

_____   _____
_____   _____
_____   _____
_____   _____
_____   _____
_____   _____

FORMAT: _____

LENS: M'F'R _____ FOCAL LENGTH ____ F-STOP ____
ATTACHMENT _____

LIGHT SOURCE _____ VOLTS ____ AMPS ____

CONDENSER SYSTEM _____

HEAT FILTER _____

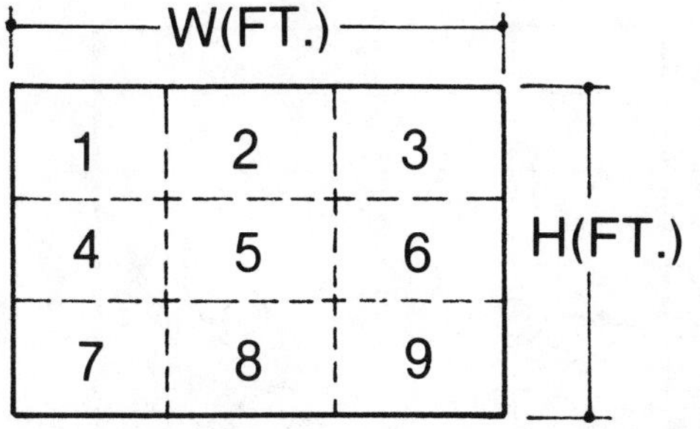

```
 ┌──────W(FT.)──────┐
 ┌─────┬─────┬─────┐
 │  1  │  2  │  3  │
 ├─────┼─────┼─────┤          H(FT.)
 │  4  │  5  │  6  │
 ├─────┼─────┼─────┤
 │  7  │  8  │  9  │
 └─────┴─────┴─────┘
```

1 ____
2 ____
3 ____
4 ____
5 ____
6 ____
7 ____
8 ____
9 ____

TOTAL LUMENS = _____

$W_x H_x FC_{AVE}$ _____

SUM = ____ ÷ 9 = AVERAGE FC = ____

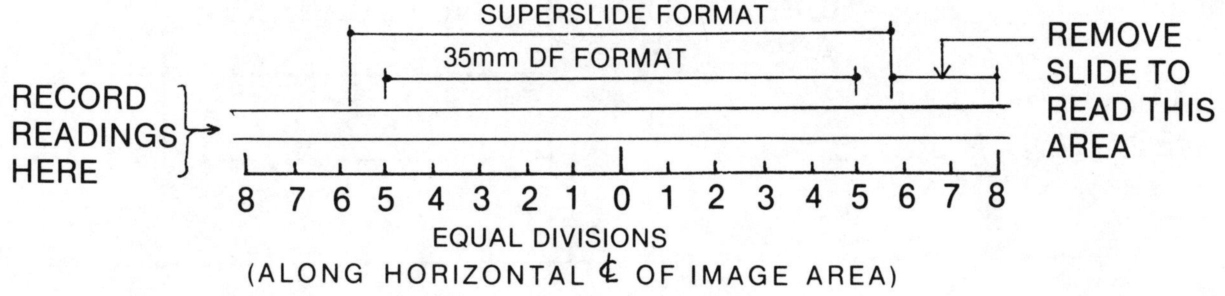

SUPERSLIDE FORMAT

35mm DF FORMAT

REMOVE SLIDE TO READ THIS AREA

RECORD READINGS HERE →

8 7 6 5 4 3 2 1 0 1 2 3 4 5 6 7 8

EQUAL DIVISIONS
(ALONG HORIZONTAL ℄ OF IMAGE AREA)

Let us redraw the concentric circles previously used to show the illumination at each of the eight divisions, recording them directly on their respective circles. We can now use this diagram as a circular graph, and record interpolated foot-candle readings lying at the center points of the nine rectangles of the ANSI grid pattern. This will be done for both the superslide and the 35-mm apertures. All necessary calculations are shown on Figs. 6-15 and 6-16.

In conclusion, we see that the ANSI 9-point method yields results that are within 1% of theoretical. A method has been proposed that divided the image area into 16 squares, in an effort to obtain more accuracy. When such a method was used, it was found to be not quite as accurate as the 9-point method. It may be concluded that the ANSI method has an accuracy well within the accuracy of the light-meter readings.

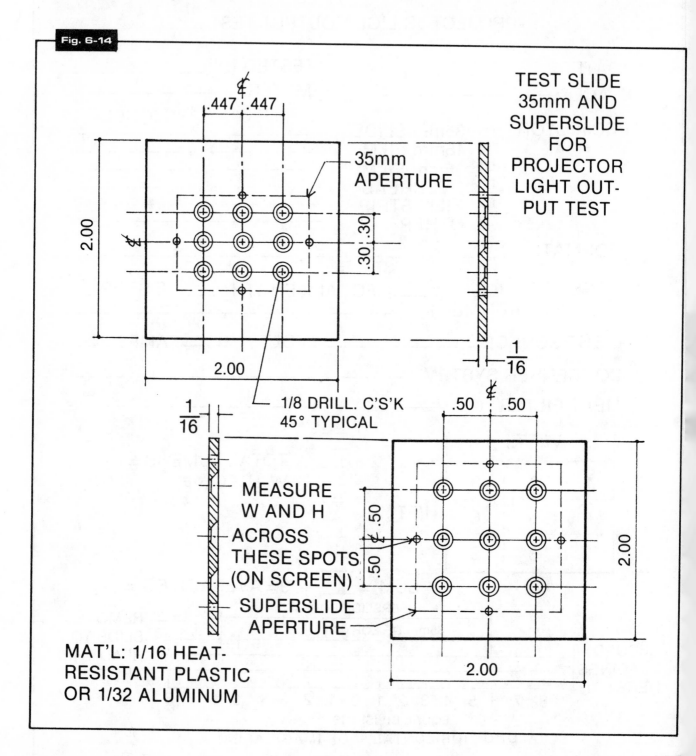

Fig. 6-14

TEST SLIDE 35mm AND SUPERSLIDE FOR PROJECTOR LIGHT OUTPUT TEST

.447 .447

35mm APERTURE

2.00

.30 .30

2.00

1/16

1/8 DRILL. C'S'K 45° TYPICAL

1/16

MEASURE W AND H ACROSS THESE SPOTS (ON SCREEN)

SUPERSLIDE APERTURE

.50 .50

.50

.50

2.00

2.00

MAT'L: 1/16 HEAT-RESISTANT PLASTIC OR 1/32 ALUMINUM

Fig. 6-15

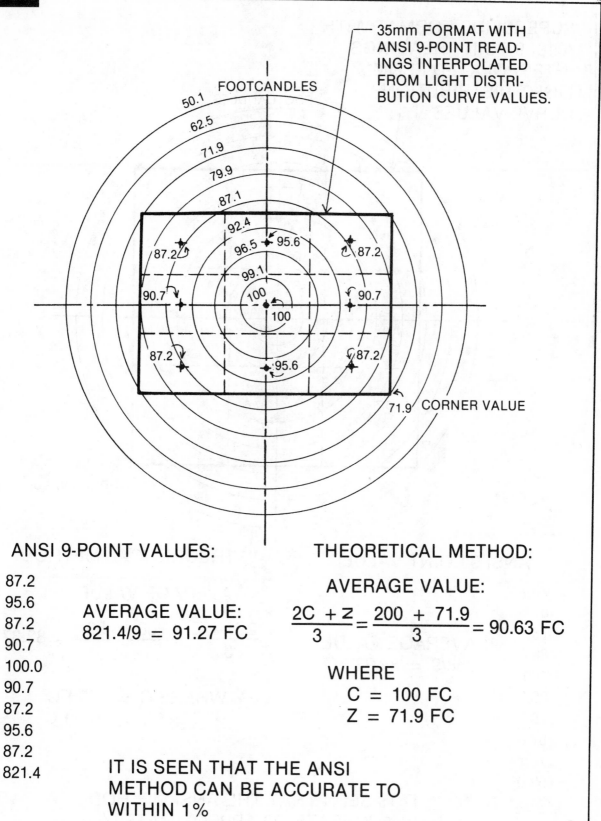

35mm FORMAT WITH ANSI 9-POINT READINGS INTERPOLATED FROM LIGHT DISTRIBUTION CURVE VALUES.

FOOTCANDLES

71.9 CORNER VALUE

ANSI 9-POINT VALUES:

87.2
95.6
87.2
90.7
100.0
90.7
87.2
95.6
87.2
821.4

AVERAGE VALUE:
821.4/9 = 91.27 FC

THEORETICAL METHOD:

AVERAGE VALUE:

$$\frac{2C + Z}{3} = \frac{200 + 71.9}{3} = 90.63 \text{ FC}$$

WHERE
C = 100 FC
Z = 71.9 FC

IT IS SEEN THAT THE ANSI METHOD CAN BE ACCURATE TO WITHIN 1%

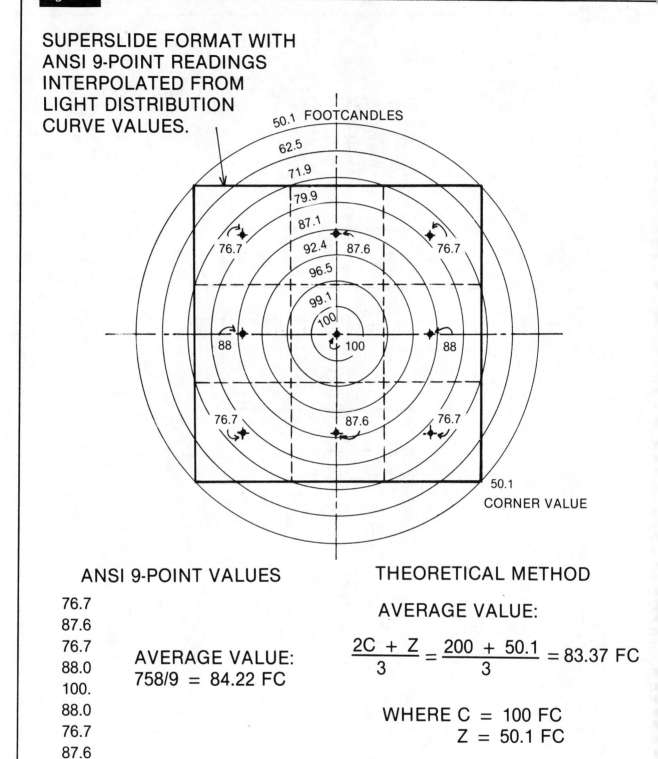

Fig. 6-16

SUPERSLIDE FORMAT WITH
ANSI 9-POINT READINGS
INTERPOLATED FROM
LIGHT DISTRIBUTION
CURVE VALUES.

50.1 FOOTCANDLES
62.5
71.9
79.9
87.1
92.4   87.6
96.5
99.1
100
100

76.7   76.7
88   88
76.7   87.6   76.7

50.1
CORNER VALUE

ANSI 9-POINT VALUES

76.7
87.6
76.7
88.0
100.
88.0
76.7
87.6
76.7
758.0

AVERAGE VALUE:
758/9 = 84.22 FC

THEORETICAL METHOD

AVERAGE VALUE:

$$\frac{2C + Z}{3} = \frac{200 + 50.1}{3} = 83.37 \text{ FC}$$

WHERE C = 100 FC
Z = 50.1 FC

IT IS SEEN THAT THE ANSI METHOD
IS ACCURATE TO APPROXIMATELY 1%

THE AUDIO IN "AUDIO-VISUAL"

There is perhaps no other segment of audio-visual technology that requires as much prerequisite knowledge and experience as that which is concerned with the design of audio systems. There seems to be a tendency for the designer to skirt the acoustical calculations and take the audio system for granted. The type of loudspeaker chosen is often the result of the designer's having found the perfect speaker — compact, medium cost, and rated by some magazine as the hi-fi speaker of the year! Its location is more a matter of where it will fit rather than where it belongs. Often, well-recognized rules-of-thumb are ignored completely. Confusion exists in differentiating between P.A. (public address) systems and voice reinforcement systems. Amplifiers are chosen with the thought that watts are cheap, and a high-power amplifier will overcome inefficient system design.

What is the reason for this apparent gap in audio system-design proficiency? The most likely reason seems to stem from the fact that sound-system engineering is not easy subject matter. It is not a compact, readily learned bit of technology. Its basics are learned on a professional engineering level, augmented by years of practical experience. Its technical roots reach into physics, acoustics, psycho-acoustics, music, speech, mathematics, electronics, and even architecture.

With the spread of technology today, it is difficult to draw lines of demarcation in professional design areas. In many cases the sound system specialist, who in past years was exactly that, is today a complete audio-visual-system contractor. Typically, such an organization started in the sound business, selling, renting, installing, and servicing all kinds of audio equipment, from those used in church fairs and high school auditoriums to the large systems needed for political rallies and football stadiums. During those years, the term sound system was not linked with audio-visual presentation, but it didn't take long before the sound-system contractor expanded his domain to include both.

Overlapping the activity of the sound-system contractor, we have his audio-visual counterpart, typified by the school audio-visual supplier, installer, and service organization. Such an organization specializes in 16-mm projector sales, rentals, and service; 16-mm films, projection screens, AV hardware, slide, filmstrip, and overhead projectors; record players, learning machines, and software packages. This type of operation is usually not heavy in sound-system engineering.

A third kind of operation has appeared in recent years, typified by both small and large contracting organizations, with the financial, and sometimes political backing to set up a successful operation in a relatively short time. These organizations do not do a retail business, but are usually franchised distributors for prominent manufacturers. Their field of operation usually includes all phases of audio-visual installation, television, and even some computer equipment installations. It is not uncommon to find that the larger organization of this kind has very competent designers and engineers on its staff, and consequently can offer complete design and installation services.

There is still another category of AV system designer/installer/supplier, and that is the vendor who manufactures his own systems and components, but also functions as a contractor, bidding on jobs engineered by others or himself. Complete engineering design services are also offered, usually strong in electronics.

Finally, there are acoustic consultants, audio-visual consultants, and communications facilities consultants. Such consultants, by the very nature of their profession, are not installation contractors or equipment suppliers. They design systems, specify their components, interconnections, and operation, and prepare system working drawings. But it is the installation contractor who takes the responsibility for supplying and installing a successful system. On his knowledge and expertise depends the success of the final system.

The more one studies the techniques of sound-system engineering, the more it is realized that much of the subject matter is inexorably tied to the field of acoustics. Sooner or later the audio-system designer must gain a working knowledge of room acoustics, including: reverberation time, absorption coefficients, sound attenuation, sound pressure level, critical distance, and a host of other pertinent subject matter that are all necessary to the design of an optimum sound system for a given space.

It seems logical that an acoustical consultant, who has collaborated with the architect to design the given space, would be eminently qualified to design the sound system for that space. And such is often the case. In certain parts of the country, it is common practice for the acoustical consultant to design the sound system as an integral part of his work. This is especially true on large projects such as music halls, theaters, coliseums, and similar spaces. The practice sometimes carries over into smaller projects, such as corporate auditoriums, lecture halls, and similar spaces, even though these spaces are more audio-visually oriented and have an audio-visual consultant designing the AV facilities. In such cases, the sound system is often handled jointly on a coordination basis.

So we see that there are all levels of knowledge represented by those who design and install sound systems for audio-visual installations. Obviously this book cannot cover all the technical prerequisites of such a wide spectrum of audio technology, for on the one hand we have a reader who needs a review of "decibles," and on the other, a reader who is interested in "the attenuation of sound in reverberant space." As stated in the preface, this is a basic text, dealing with fundamental concepts and procedures. Our discussion in this section will therefore be confined to some vital and basic concepts concerning sound pressure, sound power, and sound intensity, proceeding to a discussion of the decibel and its use in audio calculations. The remainder of the section will show how these concepts form the basis of the audio-visual system design, and permit us to investigate the behavior of sound in enclosed spaces. Space does not allow the usual descriptive discussion of loudspeakers, microphones, amplifiers, and other audio components. The available literature amply covers this material. Rather, we will stress the basic design calculations. The section concludes with typical audio system-design examples involving loudspeakers, microphones, sound levels, and amplifier power requirements.

THE CONCEPT OF *SOUND PRESSURE:* THE MICROBAR, OR DYNE PER SQUARE CENTIMETER

Sound, as depicted in Fig. 7-1, is the result of a wave motion, created by some vibrating *source,* transmitted through some *medium* (air in our case), and received by some *receptor* (the ear) acting as a transducer to assist the brain in interpreting the wave motion pressure as sound.

The air is an elastic medium and is capable of successive compression and expansion, causing extremely minute changes in the earth's static ambient air pressure in the path of a propagating sound wave. Some readers, who have not studied physics, may be surprised to learn that the weight of all the air above the surface of the earth is considerable — amounting to 14.7 pounds pressing down on every square inch of surface at sea level!

The most practical demonstration of this fact is the crude demonstration wherein a pint of water is poured into a 5-gallon thin metal container, which is left uncapped while heat is applied and the water boiled until the entire container is full of steam, thus expelling all the air within. The container is now tightly capped, and the heat source removed. If cold water is sprayed over the container, the steam inside is condensed, creating a vacuum and leaving little or no pressure inside to resist the external force of the earth's atmosphere. Result: the container is crushed as though a truck had run over it!

We can see why the human body, unlike the container, does not collapse under the weight of the atmosphere — we do not have a vacuum inside us. The atmospheric pressure is within us, and surrounds us, consequently we are not aware of this pressure at sea level. In the absence of any sound waves (pressure fluctuations), our ears would not perform their hearing function inasmuch as the air pressure on each side of the eardrum would be equalized. Any sound wave pressure encountered by the eardrum is superimposed on the steady-state atmospheric pressure, and vibration takes place in accordance with the sound pressure variation caused by the wave. The auditory mechanism of the ear and the brain produce the sensation of sound.

The most remarkable thing about the function of the human ear is the unbelievable range of sound pressures over which it operates — from the minutest pressure, corresponding to the sound of fluttering insect wings, to the pressure propagated by the deafening roar of a jet engine at take-off. This range spans a sound pressure ratio of ten million to one!

A suitable unit of pressure measurement would have to be, by definition, some unit of force acting on a unit area, such as the familiar pounds per square inch. However, the pressures we are talking about are so small that we need a much smaller unit. The Metric System offers smaller units, and is preferred throughout the scientific world. In recent years, international organizations have attempted to further consolidate and standardize the existing metric systems resulting in a system of technical measurement known as SI, from ''Systeme International d'Unites.'' This system is based on the meter, kilogram, second, ampere, Kelvin, and candela units for the basic quantities of length, mass, time, electric current, thermodynamic temperature, and luminous intensity. The reader is already faced with enough confusion in converting his thinking from the familiar English System of measurement to the Metric System, without now having to think in terms of the more recent SI System. Consequently this text will work

in terms of the more familiar CGS (centimeter, gram, second) Metric System. In any event, the starting place will be the pressure of the atmosphere at sea level and 68°F, or 20°C, as this is standard the world over.

Let us convert the 14.7 pounds per square inch pressure of the atmosphere in the English System to the metric units dynes per square centimeter, dynes/cm². To do this we need to know the following metric equivalents:

$$1 \text{ in} = 2.54 \text{ cm}$$
$$1 \text{ lb} = 453.6 \text{ grams (g)}$$
$$1 \text{ g} = 980.7 \text{ dynes}$$

The simple substitution yields

$$1 \text{ earth atmosphere} = \frac{14.7 \text{ lb/in}^2 \times 453.6 \text{ g} \times 980.7 \text{ dynes}}{(2.54 \text{ cm})^2}$$
$$= 1,013,583 \text{ dynes/cm}^2$$

This is rounded off to 1,000,000, or 1×10^6 dynes/cm², using scientific notation ($1 \times 10^6 = 1 +$ six zeros). Because typical sound pressures are extremely small, we use 1-millionth of the atmospheric pressure, or 1 dyne/cm², as the basic sound pressure measuring unit. This unit is also called 1 microbar, *micro* meaning the millionth part, and *bar* standing for barometric. Therefore 1 microbar = 1 dyne/cm². The human ear is so sensitive that it can sense sound wave pressures as low as 0.0002 dynes/cm²! This tiny pressure represents the threshold of hearing, and has become the standard sound pressure *level* against which all louder sounds are measured.

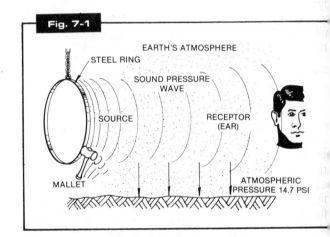

Fig. 7-1

EARTH'S ATMOSPHERE

STEEL RING

SOUND PRESSURE WAVE

SOURCE

RECEPTOR (EAR)

MALLET

ATMOSPHERIC PRESSURE 14.7 PSI

THE CONCEPT OF *SOUND POWER,* OR ACOUSTIC WORK: THE MICROWATT

In all forms of energy transfer, work is done. It is not until work is accomplished that energy is useful. The doing of work always implies that a certain amount of energy is expended in a certain time. We must remember that energy and work, as well as any other quantity, can be expressed in more ways than one, as long as they all yield the same result. The Metric System uses watts to measure all kinds of work, not just electrical power. By definition,

$$1 \text{ watt} = 10,000,000 \text{ dyne-cm/sec}$$
$$= 1 \times 10^7 \text{ dyne-cm/sec}$$

As an exercise, let us see how many watts are equivalent to 1 horsepower. (See Fig. 7-2.)

$$1\ hp = \frac{330\ lb \times 100\ ft}{1\ minute} = 33{,}000\ \text{ft-lb/minute}$$
(English system)

Converting to the Metric System,

$$1\ hp = \frac{33{,}000 \times 453.6 \times 980.7 \times 1 \times 12 \times 2.54}{60}$$

$$= 7{,}460{,}000{,}000\ \text{dyn-cm/sec. (in round numbers)}$$
$$= 746 \times 10^7\ \text{dyne-cm/sec.}$$

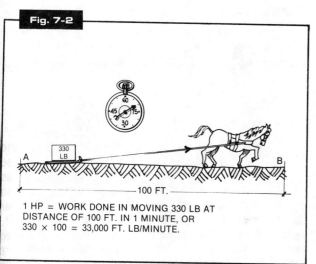

Fig. 7-2

1 HP = WORK DONE IN MOVING 330 LB AT DISTANCE OF 100 FT. IN 1 MINUTE, OR 330 × 100 = 33,000 FT. LB/MINUTE.

But 1 watt = 1×10^7 dyne-cm/sec, therefore 1 hp is equivalent to 746 watts. The single watt value of 1×10^7 dyne-cm/sec, like the atmospheric bar, is too large a unit for convenient use in sound calculations, so the microwatt (1 millionth of a watt) is used, and also the milliwatt (1 thousandth of a watt). Obviously

$$1\ microwatt = \frac{1 \times 10^7}{1 \times 10^6} = 10\ \text{dyne-cm/sec, and}$$

$$1\ milliwatt = \frac{1 \times 10^7}{1 \times 10^3} = 1 \times 10^4\ \text{dyne-cm/sec}$$

Note that when exponential terms are divided, their exponents are subtracted. Similarly, when exponential terms are multiplied, their exponents are added. The Metric System has a name for a dyne-cm, it is called an *erg*. Therefore 1 microwatt = 10 ergs/sec. The Greek letter mu, written μ, is often used to signify micro, so that we write 1 microwatt = 1 μ- watt. The microwatt is the standard unit of acoustic power.

THE CONCEPT OF *SOUND INTENSITY:* THE ERG/CM²/SEC

We have previously mentioned the erg/sec as a measure of power, or total work done per second. But now we want to see how "intense" that work is, or how much of it is performed over a specific area, such as 1 sq. centimeter. Hence the term erg/cm²/sec. The more work done over a specific area, the more intense the sound will be. The intensity of sound may be defined as the sound energy per second passing through a unit area.

If there were a point source of sound, pressure waves would radiate spherically about the source, permeating the space around it evenly in all directions. Fig. 7-3 shows how the energy is flowing radially outward from the source. But to measure how much energy is flowing in a given time, and in a

certain location, we need to examine a small volumetric section of an imaginary spherical shell. We can then see how much energy is flowing through this volume in 1 second. The reason we talk about a volume is because we have dyne-centimeters per square centimeter, and distance times area equals volume.

Actually we are talking about the amount of acoustic energy that is required to displace a unit volume of air (1 cm³) in 1 second, by virtue of the sound pressure exerted by the sound wave. Sound intensity is the measure of this acoustic energy. Its symbol is I, and its units are ergs/cm²/sec. Remember that ergs are dynes per centimeter, and the unit area we define is a square centimeter. The unit volume is 1 cm² × 1 cm in length, and the ergs/cm² are really dynes/cm/cm². We must now add a unit time limit, which is the second. Hence the intensity is expressed in ergs/cm²/sec.

We are now getting into deep water, and must be careful not to lose track of the concepts learned so far. The last remaining hurdle is to understand how the intensity, I, is evaluated. Here is the equation:

$$\text{Sound intensity} = I = \frac{p^2}{\rho c},$$

which, stated in words, says: Intensity equals the square of the sound pressure divided by the product of the air density and the velocity of sound. It is traditional to use the Greek letter rho (ρ) for air density, and also c for the velocity of sound. The symbol p represents the sound pressure. As for units,

I = ergs/cm²/sec
p = sound pressure in dynes/cm², or microbars
ρ = standard air density = 0.0012 gram/cm³, and
c = velocity of sound at sea level = 34,400 cm/sec.

But why is p squared, and why does ρc enter the equation?

While it is beyond the scope of this brief treatment to delve into the derivation of the intensity equation, we can note that it is the result of applying the principles of force, mass, acceleration, and work to air particles in a wave motion. The form of the equation reminds us of the well-known electrical expression for power, wherein $P = \frac{E^2}{R}$. The E^2 is voltage (pressure) squared, and the R is electrical resistance (impedance). In the sound intensity equation, p^2 is likewise the

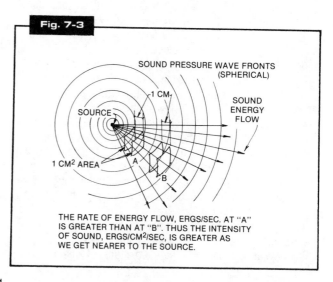

Fig. 7-3

SOUND PRESSURE WAVE FRONTS (SPHERICAL)

SOURCE

1 CM

SOUND ENERGY FLOW

1 CM² AREA

A

B

THE RATE OF ENERGY FLOW, ERGS/SEC. AT "A" IS GREATER THAN AT "B". THUS THE INTENSITY OF SOUND, ERGS/CM²/SEC, IS GREATER AS WE GET NEARER TO THE SOURCE.

acoustic pressure squared, and the term ρc is likewise called the acoustic resistance of air (impedance), or the resistance the sound pressure wave must overcome in propagating itself in air at the speed of sound. The acoustic impedance of air is, in numbers,

$$\rho c = \text{density of air} \times \text{velocity of sound}$$
$$= 0.00120 \text{ g/cm}^3 \times 34,400 \text{ cm/sec}$$
$$= 41.4 \text{ say } 42 \text{ g/cm}^2/\text{sec.}$$

This number is a constant at sea level and 20°C, and is referred to as 42 acoustic ohms.

We have just examined three very important concepts: sound pressure, sound power, and sound intensity. It will be some time before the reader who is unfamiliar with these concepts can feel comfortable with them. At this point, it would be advantageous to pause and spend a little time on the practical use of these quantities. This is more easily done, however, after reviewing the important unit of measurement used so frequently in audio work: the decibel.

A MEASURING SCALE IS NEEDED: THE *DECIBEL*

Now that we have defined sound pressure, sound power, and sound intensity, and have gotten familiar with the units in which they are expressed, we will consider the way in which our sense of hearing responds to these stimuli. The measurement of sound is aimed at approximating the reaction of the ear to sound pressure waves of varying intensities. Consequently the ratio of two sound intensities, I_2 and I_1, is expressed logarithmically, which is the way the ear evaluates different levels of sound. The unit that expresses this logarithmic ratio is the bel, named after Alexander Graham Bell. So we have

$$\text{bels} = \log_{10} \frac{I_2}{I_1}$$

It should be noted here that the human response to stimulation of our senses closely follows a logarithmic curve. This means simply that the response is approximately proportional to the logarithm of the stimulus. It is not directly proportional to the stimulus. In other words, the relationship between cause and effect is not linear, but logarithmic. Thus, a light source that is emitting twice as many light units as another, by actual measurement, does not look twice as bright to the eye. It appears to be only about ⅓ brighter (the log of 2 is 0.301). What a difficult thing to evaluate! Difficult, because it involves human reaction, which does not have numbers on it; and difficult because there are so many different reactions possible, depending on the color temperature of the light, the color of the surround, the age of the observer, and so on.

We react to sound in a similar way. When is one sound twice as loud as another? Could you identify a sound three times as loud as another? How many times louder than a sewing machine is a tractor motor? The ear is affected by pitch (frequency), loudness (amplitude), and intensity, as well as by other physical and psychological factors. So we can appreciate that measuring human response is a very subjective thing. The best we can do is accept the findings of researchers who have been concerned with the problem for the best part of the last century.

Getting back to the bel, a smaller unit was needed to be more compatible with the smallest change in intensity to which the ear can respond. Therefore a tenth of a bel was em-

ployed, and called the *decibel*, abbreviated dB. Referring to our equation for the bel, we can now write

$$\text{dB} = 10 \log_{10} \frac{I_2}{I_1}$$

We multiply the right-hand side by 10, because if the dB is a smaller unit, it requires more of them to equal the given intensity change. The decibel does not tell how much, but how many times one quantity exceeds another. It is strictly a ratio, not an absolute quantity. The decibel is therefore dimensionless.

The decibel may be used, for example, to express sound pressure ratios, for we can substitute the value of I (intensity) found earlier, in the dB expression, to get

$$\text{dB} = 10 \log_{10} \frac{\dfrac{p_2{}^2}{\rho c}}{\dfrac{p_1{}^2}{\rho c}} = 10 \log_{10} \left(\frac{p_2}{p_1}\right)^2$$

$$= 20 \log_{10} \frac{p_2}{p_1}$$

Recall that in logarithmic manipulations, the exponent 2 is brought down and multiplied by the coefficient of the log term, producing the multiplier of 20. We see that when powerlike quantities such as intensities are compared, we use the 10 multiplier, and when ratios of pressure-like quantities are involved, the multiplier becomes 20. The common logarithm is always used in these kinds of calculations, and for simplicity we will omit the base 10 subscript in any following logarithmic terms.

The decibel scale would have no meaning if it didn't have a zero starting place. Remember, the decibel compares two quantities. In the case of sound pressures, for example, when we are dealing with the sound pressure level of a single sound, there is actually also a second sound pressure, automatically taken as the least audible sound that the ear can hear. The given decibel value is then telling us that the given sound has a sound pressure level that many times as great as the threshold of sound. We learned earlier that the threshold of sound is 0.0002 dynes/cm².

dB-SPL:

If we say that the sound pressure level (hereafter referred to as SPL) of a nearby jet engine at take-off is 140 dB at 50 ft, we mean that the SPL of the engine sound is 140 times greater than the SPL of a sound that the ear can just hear. Accordingly, we must always quote the reference level when dealing with sound pressure levels in decibels, to avoid confusion. The correct terminology for the given example is 140 dB-SPL re: 0.0002 dynes/cm², at 50 ft. Table 7-1 gives some familiar sound pressure levels, listed in descending order from the loudest sound the ear can tolerate, to the faintest sound, or threshold of hearing. This is done conveniently in eight steps, each step ten times the magnitude of the preceding step. Column 3 shows that with the zero of the dB scale at the lowest audible sound pressure, all other values are greater, and are plus quantities. If we had selected some other value for the zero point on the scale, say the sound pressure due to a noisy school cafeteria at 2 dynes/cm², we end up with negative values for all lesser sounds, as shown in column 4. This is not as convenient as an all positive scale.

f we did not use the decibel scale, and just dealt with the ear sound pressures, the jet engine sound pressure would rectly be 10 million times as intense as the threshold of und, $2000 \div 0.0002 = 10,000,000$. Certainly the jet gine sound is more intense than a sound we can just hear, do we judge it to be 10 million times louder? It is obvious t the ear does not hear linearly. Let us use the dB-SPL ation to check Table 7-1 value of 60 dB-SPL for heavy auto truck traffic 100 ft from the highway.

$$dB\text{-}SPL = 20 \log \frac{p_2}{p_1}$$

ere

$p_2 = 0.2$ dyne/cm², from col. 1
$p_1 = 0.0002$ dyne/cm², from col. 1

$$dB\text{-}SPL = 20 \log \frac{0.2}{0.0002} = 20 \log 1000 = 20 \times 3$$

$$= 60 \text{ dB-SPL re: } 0.0002 \text{ dyne/cm}^2 \text{ @ 100 ft}$$

enever we talk about dB sound pressure levels, the sym- dB-SPL must be used, along with the reference level and tance. Sound pressure levels are measured on a sound el meter calibrated to read directly in decibels, referenced the threshold of sound.

ble 7.1 Typical *Sound Pressures* and Their ecibel Levels Relative to the Threshold of Sound

near Scale of Sound Pressures lynes/cm² 1	Example 2	dB-SPL re: .0002 dynes/cm² 3	dB-SPL re: 2 dynes/ cm² 4
000	Near Jet Engine at Take off (Threshold of Pain)	140	60
200	Hard Rock Band on Stage (Threshold of Feeling)	120	40
20	Loud Auto Horn at 10 feet	100	20
2	School Cafeteria with Glazed Cement Block Walls	80	0
0.2	Heavy Auto & Truck Traffic 100 Feet From Highway	60	−20
0.02	Soft Radio Music in Living Room	40	−40
0.002	Average Whisper Studio for Sound Pictures	20	−60
0.0002	Threshold of Hearing (young people, 1000 to 4000 Hz)	0	−80

-PWL

Sound power should not be confused with sound pressure. call that sound power is the capacity for doing work, and is pressed in watts in the Metric System. But the work we are king about is acoustic work. An acoustic device may draw 00 watts of electrical energy from the power line to make it erate, but because of the low acoustic efficiency of the dio device, it may generate only 10 watts of acoustic wer. This is what is meant by sound power. The measure- nt of acoustic power is not a simple, direct measurement, e sound pressure; it must be calculated from sound pres- e measurements.

Table 7-2 is similar to the sound pressure listing in Table and may be used to find the sound power of various rces and to calculate their decibel sound power levels.

Table 7-2. Typical *Sound Power* Outputs and Their Decibel Levels Relative to the Threshold of Sound

Linear Scale of Sound Power, Watts 1	Example 2	dB-PWL re: 10^{-12} Watts 3
(10^5) 100,000	Large Jet Airliner at Take off	170
(10^4) 10,000	Turbo Jet Engine 7000 Lb. Thrust	160
(10^3) 1000	Helicopter at Take off	150
(10^2) 100	4 Engine Prop. Plane at Take off	140
(10^1) 10	Symphony Orchestra Fortissimo	130
(10^0) 1	Pneumatic Chipping Hammer	120
(10^{-1}) 0.1	Automobile on Highway	110
(10^{-2}) 0.01	Ventilating Fan	100
(10^{-3}) 0.001	Loud Shout	90
(10^{-4}) 0.0001	Radio Volume Turned Up	80
(10^{-5}) 0.00001	Conversational Voice	70
(10^{-6}) 0.000001	Low Voice	60
(10^{-9}) 0.000000001	Soft Whisper	30
(10^{-12}) 0.000000000001	Threshold of Hearing	0

Here again, the linear scale represents a ratio of over its entire range of $1 \times 10^5 \div 10^{-12} = 1 \times 10^{17}$ to 1. In decibel nota- tion, this same range of power level is expressed by a range of 170 dB! This value is found as follows:

$$dB\text{-}PWL = 10 \log \frac{P_2}{P_1}$$

where

$P_2 = 100,000$ watts (from table for jetliner)
$P_1 = 0.000000000001$ watt (threshold of hearing)

$$dB\text{-}PWL = 10 \log \frac{1 \times 10^5}{1 \times 10^{-12}} = 10 \log 1 \times 10^{17}$$

$$= 10 \times 17$$

$$= 170 \text{ dB}$$

Recall that the logarithm of a number which can be expressed as 1×10^n, where n is a whole number, is simply n.

Whenever we talk about sound power levels, the symbol dB-PWL is used, along with the reference level. In American practice for some years, the reference level for zero dB was taken as 1×10^{-13} watts. International practice now uses 1×10^{-12}. If the old standard is indicated, the dB values will be 10 dB higher than for the new.

It is often convenient to simplify both the dB-SPL and dB-PWL expressions for use with the standard reference levels. We can therefore write

$$dB\text{-}SPL = 20 \log \frac{p_2}{0.0002} \quad 20 (\log p_2 - \log 0.0002)$$

$$= 20 \log p_2 - 20 (-3.69897)$$

$$= 20 \log p_2 + 74 \quad \text{re: } 0.0002 \text{ dyne/cm}^2$$

In a similar manner,

$$dB\text{-}PWL = 10 \log \frac{P_2}{1 \times 10^{-12}}$$

$$= 10 (\log P_2 - \log 1 \times 10^{-12})$$

$$= 10 \log P_2 - 10(-12)$$

$$= 10 \log P_2 + 120 \quad \text{re: } 10^{-12} \text{ watts}$$

WORKING WITH SOUND PRESSURE LEVELS OUTDOORS

In audio system design we are interested in sound pressure levels because in every AV system we have a source of sound whether it be the unaided human voice, a live orchestra, or sound that is transmitted via a loudspeaker system to the listener. Our chief interest, other than sound quality, is to be sure that the sound reaches the ear of the listener with sufficient level to permit effortless hearing.

The way in which sound level decreases as the sound travels from its source is called attenuation. Outdoors, sound attenuates differently from its behavior indoors. This is because the outdoor environment is a nonreverberant space, while indoors, the sound is reflected and absorbed by the interior surfaces, audience, and furniture.

There are two useful rules that apply when dealing with sound pressure levels and the attenuation of sound outdoors.

RULE 1

If the power input to a sound source is doubled, the SPL at a given distance will be increased by 3 dB. Conversely, if the power input to a sound source is halved, the SPL at a given distance will be decreased by 3 dB.

This relationship is apparent from the basic decibel expression for all power-like ratios:

$$dB = 10 \log \frac{P_2}{P_1}$$

If power P_2 is twice as large as P_1, then $\frac{P_2}{P_1} = 2$, and

$$dB = 10 \log 2 = 10 \times 0.301 = +3 \text{ dB (increase)}$$

If P_2 is $\frac{1}{2}P_1$, then

$$dB = 10 \log 0.5 = 10(-0.301) = -3 \text{ dB (decrease)}$$

The minus sign indicates that the sound pressure dropped.

RULE 2

If the distance from the sound source to the measuring point is doubled, the sound pressure decreases by 6 dB, or is 6 dB down (-6dB). The converse is also true.

Rule 2 involves the inverse square law, which says that for a given sound power, the sound intensity at a given distance, d_1, compared to the intensity at a greater distance d_2, will vary inversely as the square of the distance. Mathematically we can write this inverse variation as

$$\frac{I_2}{I_1} = \left(\frac{d_1}{d_2}\right)^2$$

But we have already learned that $\frac{I_2}{I_1} = \frac{(p_2)^2}{p_1}$; therefore,

from the basic decibel expression for all voltage-like (pressure) ratios, which require the 20 multiplier,

$$dB = 20 \log \frac{p_2}{p_1} = 20 \log \frac{d_1}{d_2} = 20 \log \frac{1}{2} = 20 \log 0.5$$

$$= 20(-.301) = -6 \text{ dB, or 6 dB down.}$$

A few examples showing the application of what we have learned so far in this section should help the reader feel more comfortable about working with decibels, sound intensities, sound pressure levels, sound power levels, and sound attenuation outdoors. Later in this section we will tackle typical

problems for indoor spaces, which become much more involved, because sound behaves quite differently reverberant spaces.

Example 1

The Table 7-1 of typical sound pressures given earlier shows that a loud auto horn at 10-ft distance from a listener creates a sound wave that exerts a pressure of 20 dynes/cm on the eardrum. Calculate the dB value given in column 3 the table.

Solution:

$$p_2 = 20 \text{ dynes/cm}^2, \text{ from table}$$
$$p_1 = 0.0002 \text{ dyne/cm}^2, \text{ threshold of sound}$$

$$dB\text{-SPL} = 20 \log \frac{p_2}{p_1} = 20 \log \frac{20}{0.0002}$$

$$= 20 \log (1 \times 10^5)$$
$$= 20 \times 5 = 100 \text{ Ans.}$$

We will check this answer by using the simplified equation given in the text:

$$dB\text{-SPL} = 20 \log p_2 + 74 = 20 \log 20 + 74 = (20 \times 1.3) + 74$$
$$= 26 + 74 = 100 \text{ Ans.}$$

Sometimes there is confusion between sound pressure levels and sound power levels, because both are expressed decibels. This reminds us again that the decibel is not measuring unit of some kind of sound; it does not express measurement, like pounds/sq inch or miles/hr. It tells only how many times more, or less, one quantity is compared to another, in decibels. With sound pressure levels, the decibel value compares two sound pressures, measured dynes/cm². The larger one is the given source, and the smaller one is the threshold of hearing, 0.0002 dyne/cm².

If we're talking about sound power levels, the decibel value compares two sound power levels measured in watts. The larger value is the sound power of the given source, and the smaller is the sound power output of a source at the threshold of hearing, 1×10^{-12} watt. This is 1.0 with 12 decimal places to the left of the decimal point, or 0.000000000000 watt.

Example 2

Table 7-2 of typical sound power outputs given earlier this section shows that a ventilating fan produces a sound power of 0.01 watt. Show by calculation that the sound power level agrees with the tabular value of 100 dB-PWL re: 10^{-} watt.

Solution:

$$P_2 = 0.01 \text{ watt, from table}$$
$$P_1 = 10^{-12} \text{ watt, threshold of sound}$$

$$dB\text{-PWL} = 10 \log \frac{P_2}{P_1} = 10 \log \frac{1 \times 10^{-2}}{1 \times 10^{-12}}$$

$$= 10 \log 1 \times 10^{10}$$
$$= 10 \times 10 = 100 \text{ Ans.}$$

If we use the simplified equation in the text,

$$dB\text{-PWL} = 10 \log 1 \times 10^{-2} + 120 = 10(-2) + 1$$
$$= -20 + 120 = 100 \text{ Ans.}$$

Example 3

Fig. 7-4 shows a loudspeaker on a pole that is aimed at a point B, distance of 300 feet, measured along the speaker axis. At a point A, 4 ft from the horn, a sound level meter shows an SPL reading of 128 dB re: 0.0002 dyne/cm². Find the sound pressure level at point B, the listener's location. This is a problem in inverse square law attenuation outdoors.

Solution:

Find the attentuation (drop in level) of the sound in traveling from A to B.

$$dB = 20 \log \frac{d_1}{d_2} = 20 \log \frac{300}{4} = 20 \log 75$$

$$= 20 \,(1.875) = 37.5 \text{ dB}$$

The SPL at B is then $128 - 37.5 = 90.5$ dB. Answer. If we solve this same problem by use of *RULE 2*, i.e, for every doubling of the distance we attenuate 6 dB, we can write:

    at 4 ft   SPL = 128 dB (given)
    at 8 ft   SPL = 122 dB (drop 6 dB)
    at 16 ft   SPL = 116 dB (drop 6 dB)
    at 32 ft   SPL = 110 dB (drop 6 dB)
    at 64 ft   SPL = 104 dB (drop 6 dB)
    at 128 ft SPL = 98 dB (drop 6 dB)
    at 256 ft SPL = 92 dB (drop 6 dB)
    at 512 ft SPL = 86 dB (drop 6 dB)

and by interpolation at 300 ft we get 90.5 dB. Using the second rule in this case did not save time, because there were too many steps of computation to go through. Many sound system designers use this rule exclusively for all attenuation problems, even though correct for outdoor use only.

At this point the reader may ask if an SPL of 90.5 dB at the listener location is loud enough to be heard easily. Table 7-3 is a guide to some common sound pressure levels of the human voice.

Our listener in Example 3 will evidently hear quite well under favorable ambient conditions, with a sound pressure level of 90.5 dB. It is interesting now to turn our attention to the relationship between the sound pressure level created by the loudspeaker and the watts of power that the speaker must receive from the amplifier to produce this SPL.

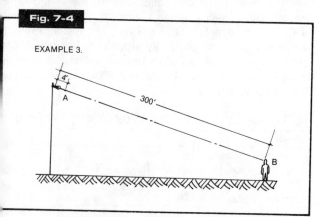

Fig. 7-4

EXAMPLE 3.

LOUDSPEAKER INPUT POWER REQUIRED

Loudspeakers are rated by the manufacturer to provide the designer with a performance criterion known as "sensitivity." It is expressed as a sound pressure level measured on axis, at a specified distance from the loud-

speaker, and with a given power input of pink noise band-limited from 500 to 3000 Hz. The standard distance is usually 4 ft, and the power input usually 1 watt, although some ratings give values at 1 meter, and sometimes at 30 ft, and sometimes at full-load power rating. Pink noise refers to the uniform filtered output of a signal generator, whereby the output is essentially flat, producing an equal energy level at all frequencies. If a pink noise filter is not used, the signal generator will produce "white noise," which contains all the frequencies perceptible to the human ear, and displays a rising characteristic of 3 dB/octave. The sole purpose of using pink noise is for the convenience of measurement.

If we know the manufacturer's sensitivity rating for a given loudspeaker, measured at 4 ft, with a 1-watt input, we can compare this with the required sound pressure level that is needed at that point, to produce the required SPL at the furthest listener location. We can then calculate how much power above the 1-watt level we need to produce the required SPL.

Example 4

Fig. 7-5 depicts a horn installation, outdoors, with a typical pole-mounted horn aimed to cover listeners up to 400-ft distance, in a certain direction. It is desired to reach the furthest listener with an SPL of 83 dB. An available horn can produce 114 dB-SPL @ 4 ft and 1-watt input. Its driver is rated to handle 100 watts of audio power. Will it do the job?

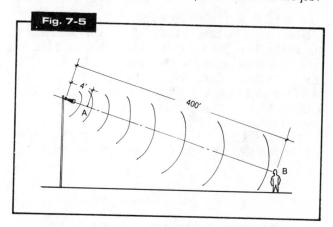

Fig. 7-5

Table 7-3. Common Sound Pressure Levels of the Human Voice

Human Voice	Sound pressure dynes/cm² @ 1-ft distance	dB-SPL @ 1-ft distance
Average whisper	0.002	20
Loud whisper	0.100	54
Normal voice	1.0	74
Loud voice	5.0	88
Shout	10.0	94

Solution:

Attenuation betweeen points A and B:

$$\text{attenuation} = 20 \log \frac{400}{4} = 20 \log 100$$

$$= 20(2) = 40 \text{ dB}$$

SPL at listener's location (given) . . . . = 83 dB

SPL required at point A . . . . = 123 dB

Note: It is customary to add 10 dB to account for the additional power delivered at the peaks of the sinusoidal audio sound waves, inasmuch as the sound pressures involved in our audio measurements are rms (root mean square) values, and not peak sine-wave values.

Total SPL at A = 123 + 10 = 133 dB
Sensitivity of given horn = 114 dB
SPL above 1 watt (133 − 114) = 19 dB

Now we have a simple case of expressing the comparison between two powers in decibel notation. We have already seen that power levels are compared in decibel notation by the relationship

$$dB\text{-}PWL = 10 \log \frac{P_1}{P_2}$$

where,

P₁ is the larger of the two powers,
P₂ is the given power, 1 watt,
dB-PWL is the power level required above 1 watt.

Substitution of the known values gives:

$$133\text{-}114 = 10 \log \frac{P_1}{1}$$

$$\frac{19}{10} = \log \frac{P_1}{1}$$

$$\log P_1 = 1.9$$

from which $P_1 = 10^{1.9} = 79.4$ watts.

Note: Raising the base 10 to the 1.9 power is called finding the antilog. Thus, the antilog of 1.9 is 79.4. Or, put another way, 79.4 is the number whose logarithm to the base 10 is 1.9. Remember the basic definition of a logarithm: a logarithm is an exponent (1.9) to which some number called the base (10) must be raised to produce the given number (79.4). Thus, $10^{1.9} = 79.4$.

The horn selection is found to be satisfactory. An amplifier with an output of 150 watts would be a good choice.

WORKING WITH SOUND PRESSURE LEVELS INDOORS

The behavior of sound in a reverberant space is a complicated process to analyze. While the mathematical derivations and physics involved in the various formulas that have been developed over the years are beyond the scope of this text, those most frequently needed will be given, and their use explained. Supporting data is given, although necessarily brief, to permit intelligent use of the information. The knowledge of logarithms, decibel notation, sound pressure levels, and so forth, is a necessary prerequisite to understanding the various processes. Typical problems will be given, along with suggested methods of solution. In an effort to keep relevant information keyed to a particular type of problem, the problem will form the text. Full pages are devoted to the problems, with calculations in the left column, and data and equations in the right, developed as the problem progresses.

Before working on the problems, some basic definitions are necessary. They are as follows:

THE DIRECT SOUND FIELD

The sound field in a room, produced by a single source such as a loudspeaker, may be thought of as divided into two parts. The first part is that portion of the field relatively near to the sound source, wherein the listener can hear the direct sound before it has been reflected from the surrounding walls, ceiling, and other surfaces. Such locations are said to be in the direct field of the sound source. A characteristic of the direct field is that the sound pressure level attenuates, or declines, as it does outdoors, dropping 6 dB for every doubling of the distance from the emitter. This is the inverse square law at work, where the attenuation is

$$20 \log \frac{d_1}{d_2} \text{ dB.}$$

THE REVERBERANT FIELD

The second part is that area which lies beyond the direct field. Here the sound pressure level drops very little, because it is now reinforced by all the reflective surfaces that the sound waves have encountered in their path of travel. As a matter of fact, the sound pressure level may remain essentially constant over large areas remote from the source.

The attenuation of sound indoors is given by the equation

$$10 \log \left[\frac{Q}{4\pi r^2} + \frac{4}{R} \right] dB,$$

where,

$\dfrac{Q}{4\pi r^2}$ term accounts for the direct radiation,

$\dfrac{4}{R}$ term accounts for the room absorption.

This equation is complicated by the inclusion of the acoustic parameters Q and R, which are defined next.

DIRECTIVITY FACTOR Q

Q is a dimensionless ratio comparing the amount of acoustic power radiated by a directional source like a horn, to that which is radiated by a nondirectional source of the same acoustic power, radiating into a spherical pattern. The value of Q naturally depends upon the type of directional device under consideration. Efficiently shaped multicellular horns have a higher Q than a cone-type loudspeaker. The human mouth forms a directive source with a rather low Q value. Typical values are given in Table 7-4.

Table 7-4. Typical Values of Directivity Factor Q

Source	Q
Person talking, no sound system used	2–2.5
Coaxial loudspeaker in infinite baffle. (Typical home hi-fi speaker)	5
Cone type woofer	5
Sectoral horns	5–9.5
Multicellular horns	5–15

The term $4\pi r^2$ is the surface area of a hypothetical sphere of radius r, the distance from the source to the point where we wish to know the attenuation.

ROOM CONSTANT R

The room constant R is a handy single-number index of the "liveness" of a given room. It depends on the acoustical treatment of the walls, ceiling, and floor, and on any other sound absorbing elements within the room such as upholstered furniture and occupants.

Each kind of wall surface, or object, has its own characteristics in the way it absorbs or reflects sound. The sound absorbing efficiency of the surface involved is given in terms of an absorption coefficient designated by the symbol α, which is the Greek letter alpha. This coefficient simply tells us what percentage of the sound striking the surface is absorbed. If a wall material has an absorption coefficient of 0.26, it means that 26% of the sound energy striking the surface is absorbed. An open window reflects no sound, all of it being absorbed by the air in the opening. Hence, its absorption coefficient is 1.0, meaning 100% of the sound is absorbed. Obviously the absorption coefficient can vary from 0 to 1.0.

Practically, sound absorption coefficients are obtained by measuring the time-rate of decay of the sound energy density in an approved reverberant room, with and without a patch of the sound absorbent material laid on the floor. Most published coefficients are found in this manner, and are referred to as α_{sab}. Sab stands for Wallace Clement Sabine (1868–1919), who discovered that the reverberation time for a room was dependent on its volume and the absorbtivity of its interior surfaces and objects.

Table 7-5 is an abbreviated table of sound absorption coefficients. Note that α varies with the frequency of the sound, but in practical calculations, the 500-Hz value is usually used.

In use, the area of each kind of surface is multiplied by its absorption coefficient. The product is expressed in sabines. In calculations involving sound fields in reverberant spaces we must find the average absorption coefficient designated as $\bar{\alpha}_{sab}$, pronounced alpha bar Sabine. This is found by summing all the products of surface S and corresponding α_{sab}, then dividing by the total surface of the interior space including walls, floor, and ceiling, and any irregular shapes.

Due to the difficulty of measuring the area of occupants, upholstered chairs, and so forth, Table 7-5 gives values in sabines per person or seat. The product of persons or seats times the sabines per person or seat may be added to the above sum of $S\alpha$, the total sum to be divided by the total S to arrive at $\bar{\alpha}_{sab}$, the average Sabine absorption coefficient.

In practical applications the room constant R may be found from

$$R = S\bar{\alpha}_{sab} \text{ ft}^2.$$

Note that R has the units of square feet, inasmuch as α is dimensionless.

REVERBERATION TIME

The reverberation time, with symbol RT_{60}, indicates the time in seconds for an initial sound in a reverberant space to drop 60 dB in sound pressure level, or to die away to 1-millionth of its original intensity.

A space with a short reverberation time, such as 0.2 second, is a "dead" space. An RT_{60} of 3 seconds would indicate an extremely live space, such as a large-volume stone church. The longer the reverberation time, the more difficult speech articulation becomes. Speech-destroying echos are associated with long reverberation times, whereas dead spaces, with short reverberation times require electronic voice enhancement to ensure good speech intelligibility. The curves in Fig. 7-6 show typical reverberation times for various space categories, versus the enclosed volume of the space.

Table 7-5. Condensed List of Sound Absorption Coefficients

Material	Frequency, Hz					
	125	250	500	1000	2000	4000
Brick, unglazed	.03	.03	.03	.04	.05	.07
Concrete block, painted	.10	.05	.06	.07	.09	.08
Glass, heavy plate	.18	.06	.04	.03	.02	.02
Glass, typical window	.35	.25	.18	.12	.07	.04
Sheetrock, ½", on studs	.29	.10	.05	.04	.07	.09
Plaster, on lath	.14	.10	.06	.05	.04	.03
Plywood, ⅜" paneling	.28	.22	.17	.09	.10	.11
Concrete block, coarse	.36	.44	.31	.29	.39	.25
Cork, 1", with air space	.14	.25	.40	.25	.34	.21
Lightweight drapes, full	.03	.04	.11	.17	.24	.35
Medium weight drapes, full	.07	.31	.49	.75	.70	.60
Heavy weight drapes, full	.14	.35	.55	.72	.70	.65
Concrete or terazzo	.01	.01	.02	.02	.02	.02
Cork, rubber pvc, on concrete	.02	.03	.03	.03	.03	.02
Wood parquet on concrete	.04	.04	.07	.06	.06	.07
Carpet, heavy, on concrete	.02	.06	.14	.37	.60	.65
Carpet, heavy, on foam rubber	.08	.24	.57	.69	.71	.73
Indoor-outdoor carpet	.01	.05	.10	.20	.45	.65
Sheetrock, ½" thick	.29	.10	.05	.04	.07	.09
Plaster, on lath	.14	.10	.06	.05	.04	.03
Acoustic tile, ¾", suspended	.76	.93	.83	.99	.99	.94
Sprayed cellulose fibers	.08	.29	.75	.98	.93	.76
Air absorption, 50% Rel. Hum. α_a = 0.9/1000 ft³						
Audience + lightly upholstered seats	sab = 4.5/seat					

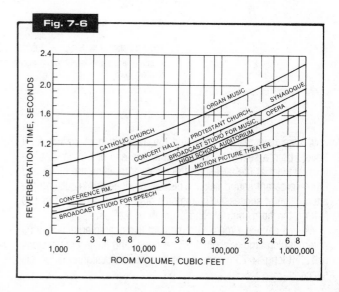

Fig. 7-6

REVERBERATION TIME, SECONDS

ROOM VOLUME, CUBIC FEET

Several equations have been developed for the calculation of reverberation time, but, for our purpose, the well-known Sabine equation will be used, where

$$RT_{60} = \frac{0.049 \times \text{volume of space}}{\text{total sabines of absorption}}$$

Let us now apply these brief acoustical procedures to an elementary problem.

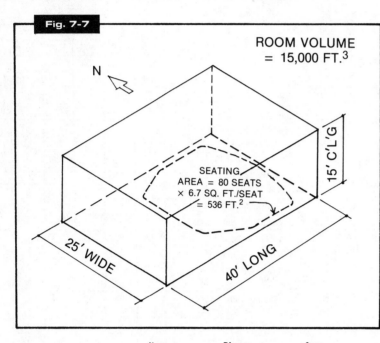

Fig. 7-7

ROOM VOLUME = 15,000 FT.^{3}

N

SEATING
AREA = 80 SEATS
× 6.7 SQ. FT./SEAT
= 536 FT.^{2}

15' C'L'G

25' WIDE

40' LONG

Example 5
ACOUSTICAL CALCULATIONS, LECTURE HALL

Applicable absorption coefficients @ 500 Hz. (from table)	
	α
N, E, W walls: plastered	.06
S wall: plywood panel	.17
Ceiling: plaster	.06
Floor: carpet, foam underlayment	.57
Audience, lightly uphol'd seats	4.5 sabines/seat
Air 50% RH α = 0.9 sabines per 1000 ft^3	

Item	Dimens.		Area		α_{sab}		sabines
N wall	40 × 15	=	600	×	.06	=	36
S wall	40 × 15	=	600	×	.17	=	102
E wall	25 × 15	=	375	×	.06	=	23
W wall	25 × 15	=	375	×	.06	=	23
Floor	40 × 25	=	1000	×	.57	=	570
Ceiling	40 × 25	=	1000	×	.06	=	60
Total surface		=	3950				
Audience & seats			80	×	4.5	=	360
Air absorption		=	$\frac{15,000}{1,000}$	×	0.9	=	14
Total sabines						=	1188

Average coefficient of absorption, $\bar{\alpha}_{sab} = \dfrac{1188}{3950} = 0.3$

Reverberation time $= \dfrac{0.049 \times V}{\text{total sabines}} = \dfrac{0.049 \times 15,000}{1188} = 0.61$ sec

Room constant $R = S\bar{\alpha}_{sab} = 3950 \times 0.3 = 1185$, say 1200 ft^2

With a reverberation time of 0.61 second, this room may be classified as a medium-to-dead room.

We are now ready to examine the attenuation of sound, as it travels from a sectoral horn source at one end of a room to a remote listener at the far end of the room, and define the direct and the reverberant sound fields. This will be done by calculating the attenuation at successive increments of distance, from 1 ft to 120 ft from the horn, and plotting a graph of the results.

We will choose a room with medium-live characteristics, with a reverberation time of about 1.75 seconds, and having dimensions of 125 ft long × 75 ft wide × 24 ft high. The design data for this space is

$V = 125 \times 75 \times 24 = 225,000$ ft^3
$S = 28,350$ ft^2
$\bar{\alpha}_{sab} = 0.22$
$RT_{60} = 1.75$ sec
$R = S\bar{\alpha}_{sab} = 28,350 \times .22 = $ approx. 6,300 ft^2

For the sectoral horn, Q = 6

Taking incremental distances, r, we can calculate the dB attenuation for each distance, using the attenuation formula given for reverberant spaces.

For r = 1 ft, attenuation = $10 \log \left[\dfrac{6}{4\pi(1)^2} + \dfrac{4}{6300} \right]$

$\qquad\qquad\qquad\qquad\qquad = -3.2$ dB

r = 2 ft $\qquad\qquad\qquad\qquad = -9.2$
r = 4 ft $\qquad\qquad\qquad\qquad = -15.2$
r = 8 ft $\qquad\qquad\qquad\qquad = -20.92$
r = 10 ft $\qquad\qquad\qquad\; = -22.67$
r = 20 ft $\qquad\qquad\qquad\; = -27.38$
r = 40 ft $\qquad\qquad\qquad\; = -30.3$
r = 80 ft $\qquad\qquad\qquad\; = -31.6$
r = 100 ft $\qquad\qquad\qquad = -31.7$
r = 120 ft $\qquad\qquad\qquad = -31.8$

The graph in Fig. 7-8 plotted on semilogarithmic graph paper (the horizontal scale is logarithmic and the vertical scale is linear), shows how the sound pressure level reaches a constant value in the reverberant field.

CRITICAL DISTANCE

If we equate the direct radiation term, $\dfrac{Q}{4\pi r^2}$, to the room absorption term, $\dfrac{4}{R}$, in the attenuation equation given above, we can solve for r, which will tell us at what distance from the sound source the direct and reverberant fields have the same attenuation. This distance is known as the critical distance, D_C. Its value will be

$$\frac{Q}{4\pi r^2} = \frac{4}{R}$$
$$16\pi r^2 = QR$$
$$r^2 = \frac{QR}{16\pi}$$

$$D_c = r = \sqrt{\frac{QR}{16\pi}} = 0.141 \sqrt{QR}$$

Fig. 7-8

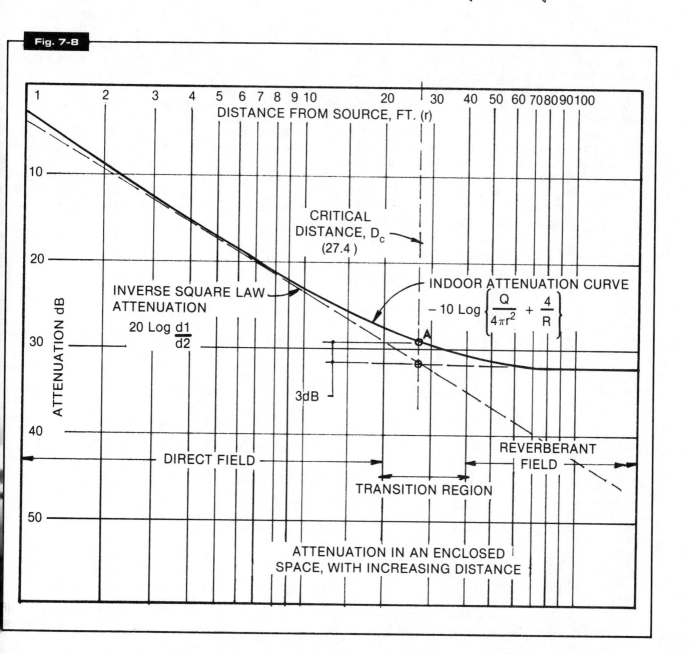

INVERSE SQUARE LAW ATTENUATION

$20 \log \dfrac{d1}{d2}$

CRITICAL DISTANCE, D_C (27.4)

INDOOR ATTENUATION CURVE

$-10 \log \left\{ \dfrac{Q}{4\pi r^2} + \dfrac{4}{R} \right\}$

ATTENUATION dB

DISTANCE FROM SOURCE, FT. (r)

3dB

DIRECT FIELD

TRANSITION REGION

REVERBERANT FIELD

ATTENUATION IN AN ENCLOSED SPACE, WITH INCREASING DISTANCE

In the preceding problem, for which we have drawn the attenuation curve, the critical distance is

$$D_C = 0.141 \sqrt{6 \times 6300} = 27.4 \text{ ft}$$

Examination of the graph shows that we can arrive at the same answer by noting that at 27.4 ft the direct, or free field attenuation (inverse square law), drops to the lowest level of the reverberant field curve. It is important to remember also that when we add two equal sound pressure levels together, the sum is 3 dB greater. In other words, when we say that the free-field sound pressure level drops to the level of the sound pressure in the reverberant field, we are combining two equal sound pressure levels, and the resultant level is 3 dB higher, as shown in Fig. 7-8 by point A on the indoor attenuation curve.

The room space between the source and the critical distance is known as the free, or direct, field, and that which lies beyond, the reverberant field. Actually, there is not a sharp line of demarcation between the two fields, and there is an area of transition lying between them. These fields are approximated on the graph.

The significance of the critical distance is first of all that it represents a sound pressure level that is within 3 dB of the maximum acoustical separation between a microphone and a loudspeaker that can ever be obtained in a given reverberant room. Once beyond the critical distance, a roving microphone would not be vulnerable to feedback. Hence the critical distance has a distinct importance where microphone/loudspeaker systems are being designed.

A workable relationship in reverberant rooms with voice amplification is that the distance from the loudspeaker to the microphone should equal or exceed the critical distance, but should not be greater than 45 ft.

In more reverberant spaces, with reverberation times in excess of 1.6 seconds, the distance from the loudspeaker to the farthest listener should not exceed about $3 D_C$. When it does, articulation loss increases. In spaces where there is no voice amplification system, the sound from a program loudspeaker can cause no feedback, but attenuation beyond the critical distance continues in accordance with the double distance/6 dB rule, while the reverberant sound pressure level remains constant, or virtually so. At twice D_C, which is 54.8 ft, there is a 6-dB drop from the indoor attenuation curve to the inverse square law attenuation line. At twice 54.8, or 109.6 ft ($4D_C$), articulation will suffer, and hearing will become difficult, due to the 12-dB drop that has accumulated by twice doubling the distance. In other words, the direct sound level has dropped 12 dB below the reverberant sound pressure level. Four times the critical distance should be the maximum distance at which any listener should be expected to hear with good intelligibility.

The following problems are typical of those encountered in practice, and are handled in sufficient detail to enable the reader to develop a methodical approach to a solution. The first problem, called Typical Problem A, concerns a centrally located speaker system, above or behind a projection screen, to transmit prerecorded sound from film sound tracks, audio tape, disks, and so forth, to an audience area in a reverberant space. Situations like this are typical of small theaters, auditoriums, lecture halls, screening rooms, and seminar rooms, where the program audio system is separate from the voice reinforcement system.

The object of the analysis is to determine the proper loudspeaker performance so that a listener in the rear of the room will hear the program material at the proper level, and without appreciable loss of articulation. The necessary amplifier wattage will also be investigated.

As we have seen in previous examples, certain acoustical parameters of the space must be known, or calculated. This information will be assumed or developed as needed.

TYPICAL PROBLEM A

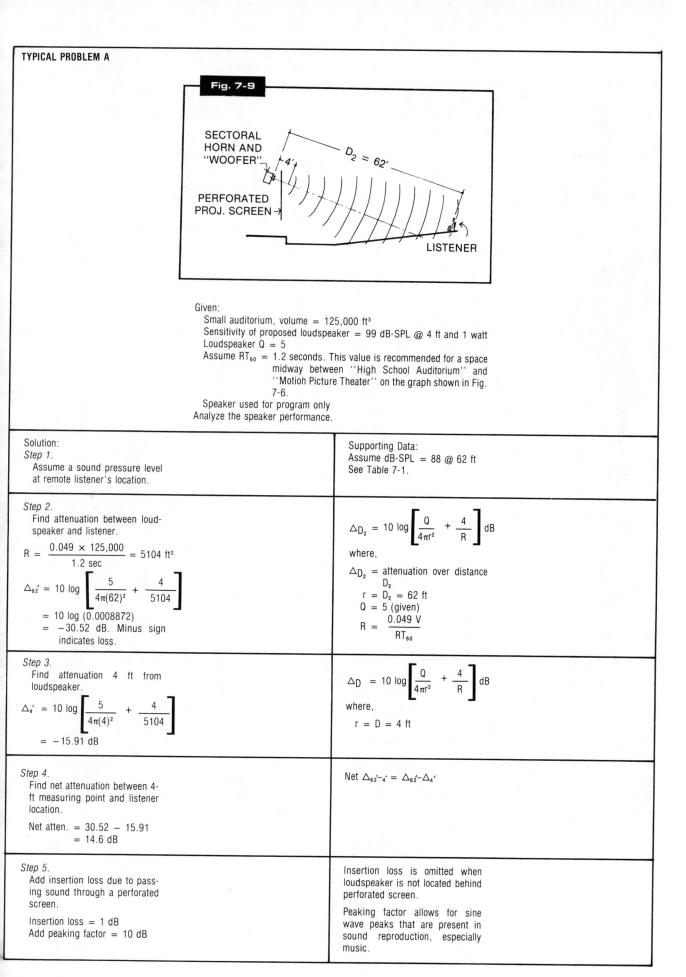

Fig. 7-9

SECTORAL HORN AND "WOOFER"

PERFORATED PROJ. SCREEN →

$D_2 = 62'$

4'

LISTENER

Given:

Small auditorium, volume = 125,000 ft³

Sensitivity of proposed loudspeaker = 99 dB-SPL @ 4 ft and 1 watt

Loudspeaker Q = 5

Assume RT_{60} = 1.2 seconds. This value is recommended for a space midway between "High School Auditorium" and "Motion Picture Theater" on the graph shown in Fig. 7-6.

Speaker used for program only

Analyze the speaker performance.

Solution:	Supporting Data:
Step 1. Assume a sound pressure level at remote listener's location.	Assume dB-SPL = 88 @ 62 ft See Table 7-1.
Step 2. Find attenuation between loudspeaker and listener. $R = \dfrac{0.049 \times 125{,}000}{1.2 \text{ sec}} = 5104 \text{ ft}^2$ $\triangle_{62'} = 10 \log \left[\dfrac{5}{4\pi(62)^2} + \dfrac{4}{5104} \right]$ $= 10 \log (0.0008872)$ $= -30.52$ dB. Minus sign indicates loss.	$\triangle_{D_2} = 10 \log \left[\dfrac{Q}{4\pi r^2} + \dfrac{4}{R} \right]$ dB where, $\triangle_{D_2}$ = attenuation over distance D_2 $r = D_2 = 62$ ft $Q = 5$ (given) $R = \dfrac{0.049 \, V}{RT_{60}}$
Step 3. Find attenuation 4 ft from loudspeaker. $\triangle_{4'} = 10 \log \left[\dfrac{5}{4\pi(4)^2} + \dfrac{4}{5104} \right]$ $= -15.91$ dB	$\triangle_D = 10 \log \left[\dfrac{Q}{4\pi r^2} + \dfrac{4}{R} \right]$ dB where, $r = D = 4$ ft
Step 4. Find net attenuation between 4-ft measuring point and listener location. Net atten. = 30.52 − 15.91 = 14.6 dB	Net $\triangle_{62'-4'} = \triangle_{62'} - \triangle_{4'}$
Step 5. Add insertion loss due to passing sound through a perforated screen. Insertion loss = 1 dB Add peaking factor = 10 dB	Insertion loss is omitted when loudspeaker is not located behind perforated screen. Peaking factor allows for sine wave peaks that are present in sound reproduction, especially music.

Step 6.
 Find operating SPL required at 4-ft standard rating distance from loudspeaker.
 	
 SPL = 14.6 + 1 + 10 + 88
 = 113.6 dB
	SPL = net attenuation +
	       insertion loss +
	       peaking factor +
	       listener level

Step 7. Find the dB above the 1-watt sensitivity produced by the loudspeaker, required to reach the value found in Step 6.	This is simply the difference between the required SPL and the rated 1-watt SPL of the loudspeaker.
Step 8. Find electric power required to produce the increase of Step 7, above 1 watt. $EPR = (10)\dfrac{14.6}{10} = 10^{1.46}$ = 28.8 watts Use a 50-watt amplifier	Let $EPR = P_1$, and 1 watt $= P_2$. Then $14.6 \text{ dB PWL} = 10 \log \dfrac{P_1}{P_2}$ $= 10 \log \dfrac{P_1}{1}$ $\log P_1 = \dfrac{14.6}{10}$ from which $P_1 = (10)\dfrac{14.6}{10}$

Step 9.
 Check that the maximum listener distance D_2 is less than $4D_c$.
 	
     $62' < 4(22.5')$
         $< 90'$ O.K.
	D_c = critical distance as found previously
	 $D_c = 0.141 \sqrt{QR}$
	 $= 0.141 \sqrt{5 \times 5{,}104}$
	 $= 22.5'$

Step 10.
 Check for minimum Q required for articulation loss of consonants not to exceed 15%.
 	
 $Q_{min} = \dfrac{641.81(62)^2(1.2)^2}{15 \times 125{,}000} = 2$
 	
 Given Q = 5 O.K.
	$Q_{min} = \dfrac{641.81(D_2)^2(RT_{60})^2}{15 \times V}$
	
	 where,
	   641.81 is a constant.
	
	     $D_2 = 62'$
	     $RT_{60} = 1.2$ sec
	       $V = 125{,}000$ ft³

The next problem, Typical Problem B, involves the use of a proscenium-mounted loudspeaker system covering the entire audience area, and used for voice reinforcement only. There is a stage microphone mounted on a lectern, and a push-to-talk microphone used in the audience area. If a program speaker were also required, it would be a separate system similar to Typical Problem A. In this application, feedback is possible because we have an open microphone on the stage which can hear the loudspeaker that is transmitting the amplified voice of the talker.

The loudspeaker should be able to transmit the amplified voice of the talker with sufficient SPL to reach the farthest listener with the same sound level that a person near the talker would experience if there were no amplification system in use.

The placement of the loudspeaker is extremely important, because its SPL, attenuated off-axis over distance D_1, may exceed the SPL of the talker's voice at the microphone. When these two sound pressure levels are equal, the condition is called *unity gain*. Beyond unity gain the system is vulnerable to feedback and howl at certain frequencies.

In this problem, we introduce the term *equivalent acoustic distance*, EAD. This is the distance, measured from the talker, at which the farthest listener thinks he is located when the sound system is in use. It varies between 3 and 20 ft, depending on how noisy the ambient space is and how well the listener hears.

TYPICAL PROBLEM B

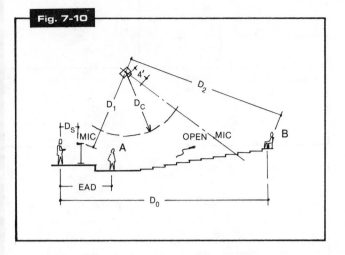

Fig. 7-10

Given distances:
$D_0 = 62'$
$D_1 = 22'$
$D_2 = 52'$
$D_s = 2'$
$EAD = 8'$

Given:
$V = 124{,}000$ ft³   Loudspeaker $Q = 5$
$S = 20{,}000$ ft²   Talker $Q = 2.5$ (human voice)
              SPL of talker $= 80$ dB @ 1' from the lips
$\bar{\alpha}_{sab} = 0.18$ Sensitivity of selected loudspeaker $= 96$ dB @ 4' and 1 watt

Analyze sound system requirements for "B" to hear with a sound system, as well as "A" does without one.

Solution:	Supporting Data:
Step 1.	$D_c = 0.141\sqrt{QS\bar{\alpha}_{sab}}$
Calculate D_c, the critical distance.	This distance should be less than D_1, the distance from loudspeaker to microphone.
$D_c = 0.141\sqrt{5 \times 20{,}000 \times .18}$	
$D_c = 18.92'$	

Step 2.	$R = S\bar{\alpha}_{sab}$ ft²
Calculate room constant R.	
$R = 20{,}000 \times 0.18$	
$R = 3600$ ft²	

Step 3.	If $\bar{\alpha}_{sab}$ is greater than 0.15, the Norris-Eyring equation is recommended.
Calculate reverberation time, in seconds.	
$RT_{60} = \dfrac{0.049 \times 124{,}000}{-20{,}000 \ln(1-0.18)}$	$TR_{60} = \dfrac{0.049\,V}{-S\ln(1-\bar{\alpha}_{sab})}$
$RT_{60} = 1.53$ seconds	where,
	ln = natural log to the base e

Step 4.	$\Delta_{8'} - \Delta_{1'} = 10 \log \left[\dfrac{Q}{4\pi r_1{}^2} + \dfrac{4}{R} \right]$
Calculate the SPL at an 8' EAD for a talker level of 80 dB @ 1' (given).	$\qquad\qquad - 10 \log \left[\dfrac{Q}{4\pi r_2{}^2} + \dfrac{4}{R} \right]$
$\Delta_{8'} - \Delta_{1'} = 23.75 - 6.99 = 16.76$	where,
SPL @ EAD $= 80 - 16.76$	$Q = 2.5$ (talker)
$\qquad\qquad = 63.24$ dB	$r_1 = 8'$
	$r_2 = 1'$
	SPL @ EAD $= 80 - (\Delta_{8'} - \Delta_{1'})$
	This is an acceptable level.

Step 5.	SPL @ "B" = SPL @ "A," a design goal. The 10-dB headroom allows for sine wave peaks, and is added to SPL @ "B."
Find "peak" level @ "B."	
SPL @ B $= 63.24 + 10$	
$\qquad\quad = 73.24$ dB	

73

Step 6. Find attenuation from 4′ measuring point, on speaker axis to farthest listener at "B". $\Delta_{52'} - \Delta_{4'} = 29.00 - 15.85$ $\qquad = 13.15$ dB loss	$\Delta_{52'} - \Delta_{4'} = 10 \log \left[\dfrac{5}{4\pi 52^2} + \dfrac{4}{3600} \right]$ $\qquad\qquad - 10 \log \left[\dfrac{5}{4\pi 4^2} + \dfrac{4}{3600} \right]$
Step 7. Find required speaker SPL @ 4′. Speaker SPL $= 73.24 + 13.15$ $\qquad\qquad = 86.39$ dB	Speaker SPL @ 4′ = SPL @ B + attenuation loss from 4′ to "B."
Note that we are now ready to compare the potential acoustic gain, PAG, with the needed acoustic gain, NAG, to determine whether or not the system is in danger of feedback. The PAG is the gain before feedback that the system can tolerate, and the NAG is the needed acoustic gain to make the farthest listener think he is as close to the talker as the EAD distance. PAG should equal or exceed NAG.	
Step 8. Find NAG. NAG $= 29.16 - 23.75 = 5.41$ dB	$\text{NAG} = \Delta_{D_0} - \Delta_{EAD}$ $10 \log \left[\dfrac{5}{4\pi 62^2} + \dfrac{4}{3600} \right] -$ $10 \log \left[\dfrac{5}{4\pi 8^2} + \dfrac{4}{3600} \right]$
Step 9. Find PAG. PAG $= \Delta_{22'} + \Delta_{62'} - \Delta_{52'} - \Delta_{2'}$ $\qquad - 6 - 10 \log 2$ $\qquad = 27.14 + 29.16 - 29 -$ $\qquad 9.97 - 6 - 3.01$ $\qquad = 8.32$ dB PAG $\geq$ NAG, therefore the system is free from feedback.	$\text{PAG} = \Delta_{D_1} + \Delta_{D_0} - \Delta_{D_2} - \Delta_{D_S}$ $\qquad - \text{FSM} - 10 \log \text{NOM}$ where, FSM = feedback stability margin, taken as 6 dB NOM = number of open microphones Each open microphone produces a drop in feedback stability.
Step 10. Calculate electric power required at the speaker input, EPR. SPL, above or below the 1-watt value given for the speaker, is $96 - 86.39 = 9.61$ dB below 1 watt $\qquad \text{EPR} = 10 - \dfrac{9.61}{10}$ $\qquad\qquad = .11$ watt	From Step 7, speaker SPL @ 4′ = 83.39 dB Sensitivity of given speaker = 96 dB @ 4′ and 1-watt input Note: the development of the EPR equation was given in Example 4 of this section.

It would appear that 0.11 watt is an extremely small amount of power, but in acoustic work we soon learn that 1 watt represents a lot of acoustic power. The average large living room hi-fi set, connected to 75–100-watt amplifiers, when turned up as loud as is comfortable, is delivering only ½ watt to the loudspeakers. And such speakers are very inefficient!

In the given problem we are asking for only 63.24 dB at the farthest listener's location. Should this be raised by 3 dB, the power requirements would double. And should the loudspeaker be less efficient, the power would increase further.

In practice, it would be normal to use a 50-watt amplifier in this application. This accomplishes two things: it guarantees a higher quality amplifier and provides ample headroom for power peaks, as well as for increased SPL at the farthest listener location.

In the last problem, typical problem C, we look at the design features of a low-level, distributed sound system using ceiling-mounted loudspeakers. Conference and meeting rooms that are small-to-medium, that is, up to about 28 ft in length, and are not overly "dead" acoustically, do not normally need a voice reinforcement system. However, when meeting rooms have to accommodate large groups of people, whether seated auditorium style, at tables, or in structured groups for seminar work sessions, it becomes difficult to hear presenters who are trying to communicate with the entire audience. Some kind of voice amplification or, more correctly, voice reinforcement, is required. This situation should not be confused with a P.A. (public address) system.

In a P.A. system, the listeners do not necessarily see the speaker; many times they are just listening to announcements. In the voice-reinforcement situation the listeners see the speaker, and are able to hear his remarks with natural sound, regardless of their listening distance.

In order to achieve uniform sound distribution, hot spots must be avoided. This requires that ample numbers of loudspeakers be used, located overhead, so that circular patterns of sound overlap at the hearing level. For optimum distribution of the full range of important frequencies, an overlap of as much as 50% in adjacent loudspeaker distribution patterns is recommended. Obviously, the lower the ceiling, the more speakers are required, and vice versa. No speaker should be expected to produce a cone of sound exceeding a 60°-apex angle where high intelligibility is desired.

While some manufacturer's data shows cone angles up to 120°, such wide dispersion is not optimum for speech. Speech intelligibility depends on uniform ear-level coverage by the higher frequencies, and these tend to beam in smaller cone angles. Where good articulation and intelligibility is not important, in background music systems, for example, the loudspeakers may be placed much further apart.

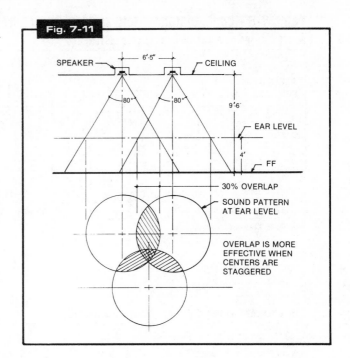

Table 7-6. Distance Between Loudspeakers (Ceiling Mounted)

50% OVERLAP* AT PLANE OF EAR (SEATED LISTENER)

LOUD-SPEAKER ANGLE OF COVERAGE	Ceiling Height, Ft													
	8'-6"	9'-0"	9'-6"	10'-0"	10'-6"	11'-0"	11'-6"	12'-0"	12'-6"	13'-0"	13'-6"	14'-0"	14'-6"	15'-0"
60°	2'-7"	2'-10"	3'-2"	3'-6"	3'-9"	4'-1"	4'-4"	4'-7"	4'-11"	5'-2"	5'-6"	5'-9"	6'-1"	6'-4"
70°	3'-2"	3'-6"	3'-10"	4'-2"	4'-7"	4'-11"	5'-3"	5'-7"	6'-0"	6'-4"	6'-8"	7'-0"	7'-4"	7'-8"
80°	3'-9"	4'-2"	4'-7"	5'-0"	5'-6"	5'-11"	6'-4"	6'-9"	7'-2"	7'-7"	8'-0"	8'-5"	8'-10"	9'-3"
90°	4'-6"	5'-0"	5'-6"	6'-0"	6'-6"	7'-0"	7'-6"	8'-0"	8'-6"	9'-0"	9'-6"	10'-0"	10'-6"	11'-0"
100°	5'-4"	6'-0"	6'-7"	7'-2"	7'-10"	8'-4"	8'-11"	9'-6"	10'-2"	10'-9"	11'-4"	11'-11"	12'-6"	13'-2"
110°	6'-5"	7'-2"	7'-10"	8'-7"	9'-3"	10'-0"	10'-9"	11'-5"	12'-2"	12'-10"	13'-7"	14'-4"	15'-0"	15'-9"
120°	7'-10"	8'-8"	9'-6"	10'-5"	11'-4"	12'-2"	13'-0"	13'-10"	14'-9"	15'-7"	16'-6"	17'-4"	18'-2"	19'-1"

*Overlap is measured on the diameter.

40% OVERLAP* AT PLANE OF EAR (SEATED LISTENER)

LOUD-SPEAKER ANGLE OF COVERAGE	Ceiling Height, Ft													
	8'-6"	9'-0"	9'-6"	10'-0"	10'-6"	11'-0"	11'-6"	12'-0"	12'-6"	13'-0"	13'-6"	14'-0"	14'-6"	15'-0"
60°	3'-1"	3'-5"	3'-10"	4'-2"	4'-6"	4'-11"	5'-2"	5'-6"	5'-11"	6'-3"	6'-7"	6'-11"	7'-3"	7'-8"
70°	3'-10"	4'-2"	4'-7"	5'-0"	5'-6"	5'-11"	6'-4"	6'-8"	7'-2"	7'-7"	8'-0"	8'-5"	8'-10"	9'-3"
80°	4'-6"	5'-0"	5'-6"	6'-0"	6'-7"	7'-1"	7'-7"	8'-1"	8'-7"	9'-1"	9'-7"	10'-1"	10'-7"	11'-1"
90°	5'-5"	6'-0"	6'-7"	7'-2"	7'-10"	8'-5"	9'-0"	9'-7"	10'-2"	10'-10"	11'-5"	12'-0"	12'-7"	13'-2"
100°	6'-5"	7'-2"	7'-11"	8'-7"	9'-5"	10'-0"	10'-8"	11'-5"	12'-2"	12'-11"	13'-7"	14'-4"	15'-0"	15'-9"
110°	7'-8"	8'-7"	9'-5"	10'-4"	11'-2"	12'-0"	12'-10"	13'-8"	14'-7"	15'-5"	16'-3"	17'-2"	18'-0"	18'-10"
120°	9'-5"	10'-5"	11'-5"	12'-6"	13'-7"	14'-7"	15'-7"	16'-7"	17'-8"	18'-8"	19'-9"	20'-9"	21'-10"	22'-11"

*Overlap is measured on the diameter.

Table 7-6—cont. Distance Between Loudspeakers (Ceiling Mounted)

30% OVERLAP* AT PLANE OF EAR (SEATED LISTENER)

LOUD-SPEAKER ANGLE OF COVERAGE	Ceiling Height, Ft													
	8'-6"	9'-0"	9'-6"	10'-0"	10'-6"	11'-0"	11'-6"	12'-0"	12'-6"	13'-0"	13'-6"	14'-0"	14'-6"	15'-0"
60°	3'-7"	4'-0"	4'-5"	4'-11"	5'-3"	5'-9"	6'-1"	6'-5"	6'-10"	7'-3"	7'-8"	8'-1"	8'-6"	8'-11"
70°	4'-5"	4'-11"	5'-4"	5'-10"	6'-5"	6'-10"	7'-4"	7'-10"	8'-4"	8'-10"	9'-4"	9'-10"	10'-3"	10'-9"
80°	5'-3"	5'-10"	6'-5"	7'-0"	7'-8"	8'-3"	8'-10"	9'-5"	10'-0"	10'-7"	11'-2"	11'-9"	12'-4"	13'-0"
90°	6'-4"	7'-0"	7'-8"	8'-5"	9'-1"	9'-10"	10'-6"	11'-2"	11'-11"	12'-6"	13'-4"	14'-0"	14'-8"	15'-5"
100°	7'-6"	8'-5"	9'-3"	10'-1"	10'-11"	11'-8"	12'-5"	13'-4"	14'-2"	15'-0"	15'-10"	16'-8"	17'-6"	18'-4"
110°	9'-0"	10'-0"	11'-0"	12'-0"	13'-0"	14'-0"	15'-0"	16'-0"	17'-0"	18'-0"	19'-0"	20'-0"	21'-0"	22'-0"
120°	11'-0"	12'-2"	13'-4"	14'-7"	15'-10"	17'-0"	18'-2"	19'-4"	20'-7"	21'-10"	23'-0"	24'-3"	25'-6"	26'-8"

Overlap is measured on the diameter.

20% OVERLAP* AT PLANE OF EAR (SEATED LISTENER)

LOUD-SPEAKER ANGLE OF COVERAGE	Ceiling Height, Ft													
	8'-6"	9'-0"	9'-6"	10'-0"	10'-6"	11'-0"	11'-6"	12'-0"	12'-6"	13'-0"	13'-6"	14'-0"	14'-6"	15'-0"
60°	4'-2"	4'-6"	5'-1"	5'-7"	6'-0"	6'-6"	6'-11"	7'-4"	7'-10"	8'-4"	8'-9"	9'-3"	9'-8"	10'-2"
70°	5'-1"	5'-7"	6'-2"	6'-8"	7'-4"	7'-11"	8'-5"	8'-11"	9'-6"	10'-0"	10'-7"	11'-2"	11'-9"	12'-4"
80°	6'-0"	6'-8"	7'-4"	8'-0"	8'-9"	9'-6"	10'-2"	10'-10"	11'-5"	12'-1"	12'-9"	13'-5"	14'-1"	14'-9"
90°	7'-2"	8'-0"	8'-10"	9'-7"	10'-5"	11'-2"	12'-0"	12'-10"	13'-7"	14'-5"	15'-2"	16'-0"	16'-10"	17'-7"
100°	8'-6"	9'-7"	10'-6"	11'-6"	12'-5"	13'-4"	14'-4"	15'-3"	16'-3"	17'-2"	18'-1"	19'-1"	20'-0"	21'-0"
110°	10'-3"	11'-5"	12'-7"	13'-8"	14'-10"	16'-0"	17'-2"	18'-4"	19'-5"	20'-6"	21'-9"	22'-10"	24'-0"	25'-2"
120°	12'-6"	13'-10"	15'-2"	16'-8"	18'-2"	19'-6"	20'-10"	22'-2"	23'-6"	24'-11"	26'-4"	27'-9"	29'-1"	30'-5"

Overlap is measured on the diameter.

The subject of loudspeakers in a ceiling, even though they are flush mounted, small in diameter (for speech), and quite unobtrusive, is certain to cause adverse comment when interior designers or architects are involved. For some unexplained reason, designers will permit air conditioning outlets, sprinkler heads, and incandescent downlights throughout the ceiling in whatever numbers necessary to give proper performance — but they will object strenuously to loudspeaker grilles cluttering the ceiling!

A ceiling speaker installation is much like an incandescent downlight system. It requires sufficient fixtures to cover the occupied space below, at the plane of the ear. Table 7-6 gives center-to-center spacing for loudspeaker placement, for various percentages of overlap, cone dispersion angle, and ceiling height. Overlap is measured on the diameter of the cone circle where it intersects the plane of the ear. Fig. 7-11 shows two adjacent ceiling speakers with 30% overlap and 80° cone pattern, mounted in a 9 ft-6 in ceiling. From Table 7-6 we find the required center-to-center distance to be 6 ft-5 in.

There are very few installations where we find as many ceiling speakers as are theoretically required. This is not because it is not desirable, but rather because someone made an arbitrary decision that the ceiling contained too many speakers, and that half as many would be satisfactory!

Typical Problem C is found in small auditoriums, lecture halls, conference rooms, and so forth, where ceiling heights are usually under 15 ft, and ceiling-mounted, low-volume-level loudspeaker systems are indicated.

It is apropos to repeat here that the foregoing material dealing with the actual audio-design parameters of typical types of spaces is not offered as an absolute methodology. Indeed, there is considerable room for controversy among designers, acousticians, and engineers. The best that can be hoped for is that the reader will use the information presented, in conjunction with his own experience and knowledge, to further his understanding of the processes involved.

The subject matter discussed has barely scratched the surface. The reader can recognize the complexity of the technology and realize the need for further study and investigation. To this end, a brief bibliography follows, in the hope that each interested reader will pursue selected material and put together methods and procedures to assist him or her in the audio design work.

TYPICAL PROBLEM C

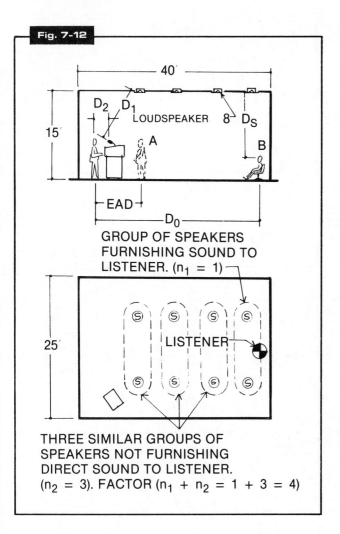

Fig. 7-12

40′

15′

D_2 D_1
LOUDSPEAKER 8↧ D_S
A
B

EAD

D_0

GROUP OF SPEAKERS
FURNISHING SOUND TO
LISTENER. ($n_1 = 1$)

25′

LISTENER

THREE SIMILAR GROUPS OF
SPEAKERS NOT FURNISHING
DIRECT SOUND TO LISTENER.
($n_2 = 3$). FACTOR ($n_1 + n_2 = 1 + 3 = 4$)

Given data:
Use acoustical data from example
5.
$V = 40′ \times 25′ \times 15′ = 15,000$ ft^3
$Q = 5$ for ceiling speakers
$Q = 2.5$ for talker
$R = 1200$ ft^2
$RT_{60} = 0.61$ sec
No. speakers = 8
Sensitivity = 88 dB-SPL at
 4′ and 1 watt
EAD = 4 ft = r_1
$D_S = 1′ = r_2$
$D_1 = 9′ - 0″$
$D_2 = 11′ - 0″ = r$
$D_0 = 35′ - 0″ = r_3$

Analyze loudspeaker selection and find electric power required.

Solution:
Step 1.
 Calculate D_C, the critical distance.

$$D_C = 0.141 \sqrt{\frac{5 \times 1200}{4}} = 5.46′$$
$$= 5′\text{-}6″$$

Microphone is not in the direct field of loudspeaker.

Supporting Data:
D_C for a distributed sound system
may be evaluated from

$$D_C = 0.141 \sqrt{\frac{QR}{(n_1 + n_2)}}$$

Factor ($n_1 + n_2$) is explained in
Fig. 7-12.

NOTE: The open microphone on the lectern is out of the direct field of the nearest loudspeaker overhead, and should not be vulnerable to feedback. Nonetheless, the loudspeaker nearest the lectern should be separately circuited so that it may be shut off if necessary, when the lectern microphone is in use.

Obviously the factor ($n_1 + n_2$) allows for a reduction in system gain when a distributed system is used, hence shortening D_C.

Step 2. Establish an equivalent acoustic distance, EAD, and calculate the dB-SPL at this point (point "A"). Assuming the talker produces 76 dB-SPL at the microphone (D_S = 1') at EAD $SPL = \Delta_4' - \Delta_1'$ $\quad = 18.02 - 6.94$ $\quad = 11.06$ dB $SPL @ EAD = 76 - 11.06 = 64.94$ $\qquad$ say 65 dB-SPL	The EAD should be chosen so that a satisfactory sound level exists at this point, consistent with the noise level of the space. A 4' EAD will be assumed. Remember that Δ_4' means $$10 \log\left[\frac{Q}{4\pi r_1{}^2} + \frac{4}{R}\right] - \text{etc.}$$ where, $\quad Q = 2.5$ (voice) $\quad r_1 = 4$ $\quad r_2 = 1 = D_S$ $\quad R = 1200$
Step 3. Find peak level at "B," the listener's position. The same SPL will be available at "B" as at EAD, by definition. dB-SPL @ "B" = 65 + 10 = 75	Add 10 dB for "head room" on sine-wave peaks.
Step 4. Find attenuation from 4' measuring point in front of loudspeaker to furthest listener "B." $\Delta_{11}' - \Delta_4' = 21.79 - 15.5$ $\qquad\qquad = 6.29$ dB-SPL	Attenuation = $\Delta_{11}' - \Delta_4'$ *using* $\quad Q = 5$ $\quad R = 1200$ $\quad r = 11' = D_2$ $\quad r_1 = 4'$
Step 5. Find required SPL @ 4' Loudspeaker SPL = 75 + 6.29 $\qquad\qquad\qquad = 81.29$ dB-SPL	Speaker SPL = SPL @ "B" plus attenuation from 4' to "B."
Step 6. Find needed acoustic gain, NAG. $\quad$ NAG = 24.37 - 15.50 $\qquad\quad = 8.87$ dB	$NAG = \Delta_{D_0} - \Delta_{EAD}$ $\quad$ using Q = 5 $\qquad\quad r_3 = D_0 = 35'$ $\qquad\quad r_1 = 4'$
Step 7. Find potential acoustic gain, PAG. PAG = 20.84 + 24.37 $\qquad\quad - 21.79 - 3.97 - 6$ $\qquad = 13.45$ dB PAG $\geq$ NAG, therefore there should be no feedback problem.	$PAG = \Delta_{D_1} + \Delta_{D_0} + \Delta_{-\Delta_{D_2}} - \Delta_{D_S}$ $\qquad\quad - \text{FSM} - 10 \log \text{NOM.}$ using r $\quad = D_2 = 11'$ $\qquad r_1 \quad = 4'$ $\qquad r_2 \quad = D_S = 1'$ $\qquad r_3 \quad = D_0 = 35'$ $\qquad Q \quad = 5$ $\qquad R \quad = 1200$ $\qquad \text{FSM} = 6$ dB $\qquad \text{NOM} = 1$
Step 8. Calculate electric power required SPL below 1 watt = 88 - 81.29 = 6.71 $EPR = (10)^{\frac{-6.71}{10}} = 0.21$ watt/speaker Total watts = 8 × 0.21 = 1.68 Use a 50-watt amplifier.	Using sensitivity given as 88 dB-SPL @ 4' and 1 watt, and required speaker SPL from Step 5. Use ½-watt tap on speaker transformer. As discussed following Typical Problem "B."

For Further Reading

Beranek, L. L., *Noise and Vibration Control*. New York: McGraw-Hill Book Company, 1971

Chiswell, B. and Grigg, E., *S. I. Units*. Sydney: John Wiley & Sons Australasia Pty Ltd., 1971

Clifford, M., *Microphones — How They Work & How to Use Them*. Blue Ridge Summit, Pa: Tab Books, 1979

Cohen, A. B., *Hi-Fi Loudspeakers and Enclosures*. Rochelle Park, N.J.: Hayden Book Company, Inc., 1978

Davis, D. and C., *Sound System Engineering*. Indianapolis: Howard W. Sams & Company, Inc., 1977

Doelle, L. L., *Environmental Acoustics*. New York: McGraw-Hill Book Company, 1972

Egan, M. D., *Concepts in Architectural Acoustics*. New York: McGraw-Hill Book Company, 1972

Rettinger, M., *Acoustic Design and Noise Control*, Vol. 1. New York: Chemical Publishing Company, 1977

Tremaine, H., *Audio Cyclopedia*. Indianapolis: Howard W. Sams & Company, Inc., 1969

Wood, A., *The Physics of Music*. New York: Dover Publishers, Inc., 1961

KEYSTONED IMAGES

Most viewers are familiar with keystoned images as they relate to the overhead projector, particularly vertical keystoning, which results when we project at an upward angle on a vertical screen from a table or cart-mounted projector. While it is common practice for presenters and audiences alike to ignore rather appreciable amounts of keystone when using the overhead projector, such distortion is unacceptable in the presentation of slides, especially when multiple image configurations are used.

Motion picture theaters often have projection booths located above the balcony level at the rear of the house. This is especially true of the older theaters in the larger cities which were formerly legitimate playhouses before being converted to cinemas. It is not uncommon to find projector downtilt angles up to 25° in these older houses. Such a downtilt angle would obviously produce an unacceptable trapezoidal format on the screen were it not for the fact that the aperture plates in professional 35-mm projectors are filed to compensate for this distortion. However, this compensation corrects only the sides of the trapezoidal image, making them rectangular. The eye is satisfied with the rectilinear appearance of the image format, even though the content of the image is still distorted. Also, the eye has a tendency to be less critical of a picture when it is in motion.

Slide images, on the other hand, are static, and often are not pictorial, containing chart lines, graphs, tabular information, etc. Rectilinearity, then, becomes something of prime importance for both image content and aperture configuration.

WHAT KINDS OF KEYSTONE ARE TO BE AVOIDED IN SLIDE PROJECTION SYSTEMS?

We encounter three kinds of keystone — vertical, horizontal, and a combination of both. We will concern ourselves with the keystoned images of 35-mm slides, but the analysis techniques developed may also be applied to the projection of other formats. Fig. 8-1 shows all three kinds of keystone as they might appear in a single, and dual-image configuration. A flat, vertical screen is assumed, and the amount of keystone has been exaggerated for clarity.

Fig. 8-1

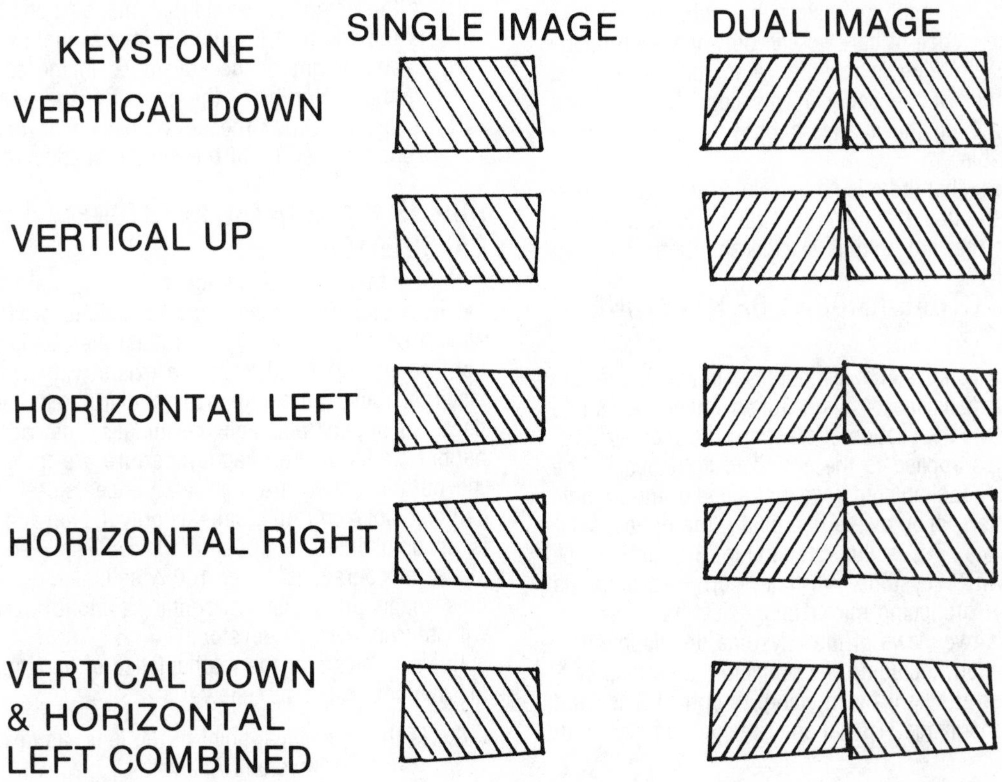

KEYSTONE	SINGLE IMAGE	DUAL IMAGE
VERTICAL DOWN		
VERTICAL UP		
HORIZONTAL LEFT		
HORIZONTAL RIGHT		
VERTICAL DOWN & HORIZONTAL LEFT COMBINED		

IS THERE ANY WAY OF CORRECTING A LENS TO COMPENSATE FOR KEYSTONING?

Unfortunately, no. The only solution to the keystone problem is to avoid it, or keep it within acceptable limits.

IS THERE ANY ANGULAR LIMIT ON THE PROJECTION AXIS BEYOND WHICH THE RESULTING KEYSTONE IS UNACCEPTABLE?

Some investigators have concluded that a projection axis angle not exceeding 5 or 6° is acceptable on a vertical screen. Such a rule of thumb makes no allowance for whether the image is front, or rear projected, or whether the skew angle is vertical or horizontal. Five degrees of projector tilt on a long throw front-projected image causes less distortion than the same 5° on a short throw rear-projected image.

The better criterion for acceptable keystone is percent distortion rather than projection axis angle. The percent distortion in any projected format is found by comparing the length of the two opposite sides of the image that are parallel, but of different lengths, and finding the percentage that the longer exceeds the shorter. An example is worked out in Fig. 8-2.

When a single image is projected on the screen, a 2% difference between the lengths of opposite parallel sides is acceptable. This is true for horizontal as well as vertical keystone. But when dual or triple images are used with vertical keystoning, this same amount of distortion may not be considered acceptable, due to the appearance of noticeable triangles of bare screen showing between adjacent images.

As an example, if we had three adjacent vertically oriented slides, each normally 50-in wide by 74-in high, projected at a downtilt of 5°, from a rear screen projector placed 18 ft-6 in from the screen, the resulting distortion would be just less than 2%, but the wedge of bare screen between each image would measure 1-in wide at the top when the images just touched each other at the bottom. This condition could be helped by overlapping the images slightly, but the overlap would be noticeable.

We can conclude that there is no hard and fast rule for keystone limits, as the final decision must take into account the several variables that are unique to each situation.

HOW IS THE ACTUAL AMOUNT OF KEYSTONE CALCULATED?

The calculation of keystone, or more specifically, the calculation of the dimensions of a keystoned image, are quite readily found by the application of standard trigonometric triangle relationships applied to the projected light beam. The process is somewhat simplified if we make use of the triangle formed by the intersection of the film plane, the optical axis, and the image plane. This is illustrated in Fig. 8-3 for the case of vertical ''down'' keystone. Vertical ''up'' keystone is handled similarly, producing mirror image results.

Fig. 8-3 shows two views of the keystone development:

A. a side elevation, including the projector, a vertical section taken through the light beam along its optical axis, and an edge view of the image, TB, taken at the vertical center of the screen,

B. a front or rear view of the keystoned image, compared to the size of a normal image (dashed line) at the same throw distance, 0 ft-0 in.

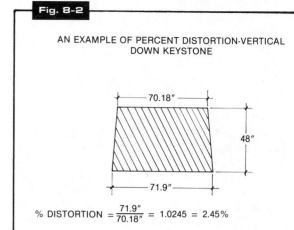

Fig. 8-2

AN EXAMPLE OF PERCENT DISTORTION-VERTICAL DOWN KEYSTONE

$$\% \text{ DISTORTION} = \frac{71.9''}{70.18''} = 1.0245 = 2.45\%$$

In Fig. 8-3:

0′ is the front focal point of the lens, from which the throw distance TD has been measured. Note: as discussed in the earlier section on optics, the throw distance actually includes the front focus distance, and TD is the distance ''x'' in the lens formula. We have simply referred to x as TD because it is more convenient, and the accuracy of the analysis is not affected.

Θ is the angle of downtilt, degrees, which also equals the angle formed by the vertical centerline of the screen and the line $0' - P_v$ parallel to the film plane.

β is the half-angle of the projector light beam, degrees.

φ is half the included angle between the sides of the keystoned image, which meet at P_v.

H is the height of the normal image, inches.

W is the width of the normal image, inches.

H_t is the height of the keystoned image *above* the horizontal centerline of the normal image, inches.

H_b is the height of the keystoned image *below* the horizontal centerline of the normal image, inches.

HOW IS HORIZONTAL KEYSTONE CALCULATED?

Horizontal keystone is handled in a manner similar to vertical keystone insofar as the calculations are concerned. It should be noted, however, as illustrated earlier in this section, when horizontal keystone exists with two or more adjacent images, the pie-shaped angular patch of vacant screen does not appear between the images. Instead, the top and bottom edges of the imaged aperture are trapezoidal — but the butting edges are vertical. Hence, side-by-side images can be displayed rather satisfactorily if a top and bottom mask is used to hide the upper- and lower-splayed edges. Keystone, of course, is never 100% satisfactory — but in multiple-image projection horizontal keystone is easier to deal with than is vertical keystone.

In the analysis of the mathematics of horizontal keystone the symbols list is as follows:

0′ is the front focal point of the lens, as described in the ''Vertical Keystone Analysis.''

Θ is the angle of skew, degrees, which also equals the angle formed by the horizontal centerline through the focal point, 0′F, and the plane of the screen.

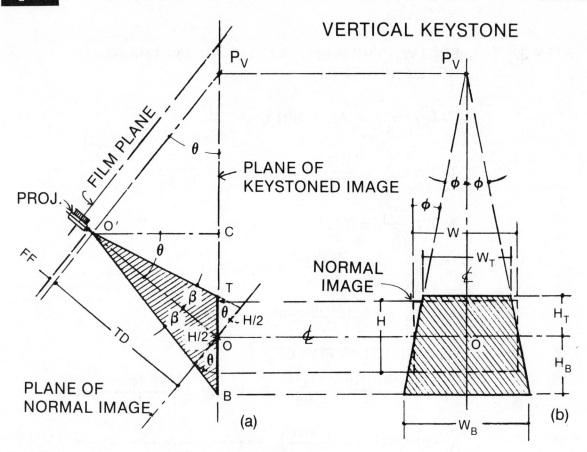

Fig. 8-3

VERTICAL KEYSTONE

(a)

(b)

STEP 1.   IN $\triangle$ O'TO, FIND OT: ($= H_T$)

$$O'O = T.D. = \frac{H}{h} \times f \text{ OR } \frac{W}{w} \times f$$

$$\beta = TAN^{-1}\frac{\frac{H}{2}}{TD} = TAN^{-1}\frac{H}{2TD}$$

$$\angle O'OT = (90 - \theta)$$
$$\angle O'TO = 180 - (\beta + 90 - \theta) = (90 - \beta + \theta)$$
FROM THE LAW OF SINES:

$$\frac{OT}{SIN\ \beta} = \frac{T.D.}{SIN(90 - \beta + \theta)}$$

FROM WHICH

$$\boxed{H_T} = OT = \frac{T.D.(SIN\ \beta)}{SIN\ (90 - \beta + \theta)} \quad ————————(1)$$

STEP 2.

IN A SIMILAR MANNER, FIND $H_B = OB$

$$\boxed{H_B} = OB = \frac{T.D.\ (SIN\ \beta)}{SIN\ (90 - \beta - \theta)} \quad ————————(2)$$

Fig. 8-3 CONT'D

STEP 3. FIND W_T, THE WIDTH AT THE TOP OF THE IMAGE. NOTE FIRST THAT:

$$\text{TAN } \phi = \frac{\frac{W}{2}}{OP_v} \quad \text{AND} \quad \text{SIN } \theta = \frac{T.D.}{OP_v}$$

THEN $OP_v = \dfrac{T.D.}{SIN\theta}$ AND TAN $\phi = \dfrac{WSIN\theta}{2\,T.D.}$

NOW $\dfrac{W_T}{2} = \text{TAN } \phi\,(OP_v - H_T)$

$$= \frac{WSIN\,\theta}{2\,T.D.}\,(OP_v - H_T)$$

OR $W_T = \dfrac{2WSIN\theta}{2\,T.D.}\,(OP_v - H_T)$

SUBSTITUTING FOR OP_v:

$$W_T = \frac{WSIN\theta}{T.D.}\left(\frac{T.D.}{SIN\theta} - H_T\right) = W - \frac{WSIN\theta H_T}{T.D.}$$

$$\boxed{W_T} = W\left(1 - \frac{H_T SIN\theta}{T.D.}\right) \quad \text{-----------} \quad (3)$$

STEP 4. IN A SIMILAR MANNER, FIND W_B, THE WIDTH AT THE BOTTOM OF THE IMAGE.

$$\boxed{W_B} = W\left(1 + \frac{H_B SIN\theta}{T.D.}\right) \quad \text{--------------} \quad (4)$$

RECAP.

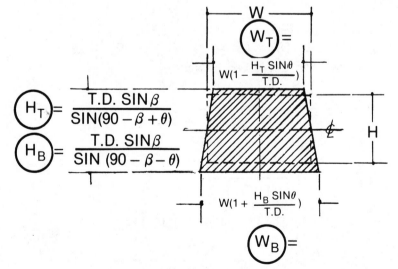

$$H_T = \frac{T.D.\ SIN\beta}{SIN(90 - \beta + \theta)}$$

$$H_B = \frac{T.D.\ SIN\beta}{SIN(90 - \beta - \theta)}$$

β is the half-angle of the projector light beam, as before.

ϕ is half the included angle between the top and bottom edges of the keystoned image, which meet at P_h, degrees.

H is the height of the normal image, inches.

W is the width of the normal image, inches.

H_L is the height of the keystoned image on the short side, here called the left side, inches.

H_R is the height of the keystoned image long side, here called the right side, inches.

W_L is the width from the centerline to the short side of the keystoned image, inches.

W_R is the width from the centerline to the long side of the keystoned image, inches.

Fig. 8-4 shows two views of the development:

(a) a plan view including the projector, a horizontal section taken through the light beam along its optical axis, and an edge view LR of the image taken at the horizontal center of the screen,

(b) a front or rear view of the keystoned image, compared to the size of the normal image (dashed line) at the same throw distance, 0'-0.

IS THERE ANY WAY TO AVOID USING THE CUMBERSOME EQUATIONS JUST DEVELOPED TO ARRIVE AT THE AMOUNT OF KEYSTONE?

Yes. If the previously presented equations are solved for given angles of light beam angularity and projector throw distance, a table of constants can be calculated in terms of the standard width of an image projected normal to the screen. These calculations have been made for 35-mm slides, and appear in Tables 8-1 and 8-2. The same could be done for any other format. The first table is used for vertical keystone, and the second for horizontal.

The heavy line running diagonally through the tables is drawn at a point representing not more than 2% distortion. All combinations of throw distance and angle of tilt, or skew, which yield tabular values to the right of the heavy line, represent "safe" conditions wherein the amount of distortion does not exceed 2%.

The following example will not only demonstrate the use of the keystone tables, but will also provide a review of some important procedures previously covered.

Example 1.

Calculate the keystoned dimensions resulting when a correctly sized 35-mm slide is projected at a downtilt of 5°, on a vertical screen, at a throw distance of 48 ft measured to the center of the screen image. The distance to the farthest viewer is 52 ft.

Solution

Optimum height of the image is ⅛ of the farthest viewer distance, or 52 ÷ 8 = 6.5 ft = 6 ft-6 in. The corresponding width of the image is 1.48 × 6.5 = 9.62 ft or 9 ft-7 in. Note that while this image size is desirable, we are not certain that we will be able to produce it with one of the standard lenses available. Consequently, we must calculate the required focal length of the lens. Using the standard lens formula, T.D. = (M), we recall that magnification M is the same as image height ÷ aperture height, or $\dfrac{H}{h}$. Thus:

$$\frac{6.5'}{0.902''} \times f'' = 48'', \text{ from which } f = 6.66''$$

The nearest standard focal length lens is 7 in, and a corrected image height must now be found by using the preceding equation once more.

$$\frac{H'}{0.902''} \times 7'' = 48',$$

from which H = 6.19' = 74.28"
and W = 9.16' = 109.92"

These two dimensions represent the normal image, without keystone, with a standard available lens and a 48-ft throw distance.

From the table of vertical keystone constants on the following pages, record values listed for a throw distance of 48 ÷ 9.16 ft = 5.24 W. It is accurate enough to use the column for T.D. = 5 W. The four values listed must be multiplied by W, the normal width = 109.92 in as shown:

H_T = .3371 × 109.92 = 37.05"
H_B = .3411 × 109.92 = 37.49"
W_T = .9941 × 109.92 = 109.27"
W_B = 1.0059 × 109.92 = 110.57"

The tabulated constants were found to the right of the heavy line, therefore the distortion is less than 2%. The actual distortion is:

$$1 - \left[\frac{109.27}{110.57}\right] 100 = 1.18\%$$

Table 8-1.
VERTICAL Keystone Constants

35-mm 2×2 Slides

ANGLE OF DOWNTILT & ITEM		THROW DISTANCE (IN TERMS OF ST'D IMAGE W)						
		1W	2W	3W	4W	5W	6W	7W
NOTE: ALL VALUES IN THIS TABLE MUST BE MULTIPLIED BY W, THE WIDTH OF THE NORMAL IMAGE.								
1°	H_T	.3359	.3369	.3372	.3374	.3375	.3376	.3376
	H_B	.3399	.3389	.3386	.3384	.3383	.3382	.3382
	W_T	.9941	.9971	.9980	.9985	.9988	.9990	.9992
	W_B	1.0058	1.0030	1.0020	1.0015	1.0012	1.0010	1.0008
2°	H_T	.3341	.3361	.3367	.3371	.3372	.3374	.3375
	H_B	.3421	.3401	.3394	.3390	.3388	.3387	.3386
	W_T	.9883	.9941	.9961	.9971	.9976	.9980	.9983
	W_B	1.0119	1.0059	1.0039	1.0030	1.0024	1.0020	1.0017
3°	H_T	.3324	.3353	.3363	.3368	.3371	.3373	.3374
	H_B	.3445	.3413	.3403	.3398	.3395	.3393	.3392
	W_T	.9826	.9912	.9941	.9956	.9965	.9971	.9975
	W_B	1.0180	1.0089	1.0059	1.0044	1.0036	1.0030	1.0025
4°	H_T	.3308	.3347	.3360	.3367	.3371	.3373	.3375
	H_B	.3470	.3427	.3414	.3407	.3403	.3400	.3398
	W_T	.9769	.9883	.9922	.9941	.9953	.9961	.9966
	W_B	1.0242	1.0120	1.0079	1.0059	1.0047	1.0040	1.0034
5°	H_T	.3294	.3342	.3358	.3366	.3371	.3375	.3377
	H_B	.3495	.3442	.3425	.3417	.3411	.3408	.3406
	W_T	.9713	.9854	.9902	.9927	.9941	.9951	.9958
	W_B	1.0305	1.0146	1.0100	1.0074	1.0059	1.0050	1.0042

Fig. 8-4

HORIZONTAL KEYSTONE

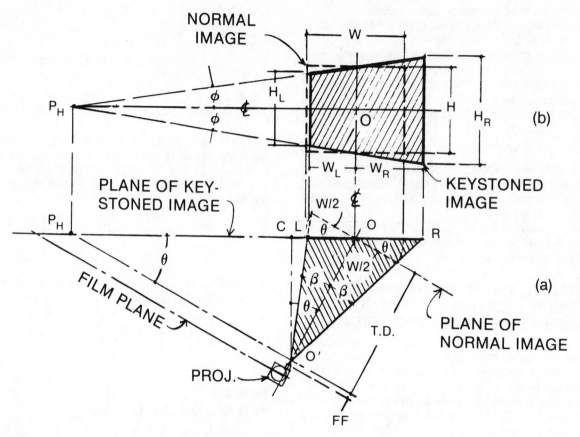

STEP 1.   IN $\triangle$ O'LO FIND OL: ($=W_L$)

$$O'O = T.D. = \frac{H}{h} \times f \ OR \ \frac{W}{w} \times f$$

$$\beta = TAN^{-1} \frac{\frac{W}{2}}{T.D.} = TAN^{-1} \frac{W}{2 \ T.D.}$$

$\angle O'OL = (90 - \theta)$
$\angle O'LO = 180 - (B + 90 - \theta) = (90 - \beta + \theta)$

FROM THE LAW OF SINES:

$$\frac{OL}{SIN \beta} = \frac{T.D.}{SIN (90 - \beta + \theta)}$$

FROM WHICH

$$\boxed{W_L} = OL = \frac{T.D. \ SIN \ \beta}{SIN (90 - \beta + \theta)} \ \text{-------------- (5)}$$

STEP 2.   IN A SIMILAR MANNER, FIND W_R = OR

$$\boxed{W_R} = OR = \frac{T.D. \ SIN \ \beta}{SIN (90 - \beta - \theta)} \ \text{-------------- (6)}$$

Fig. 8-4 CONT'D

STEP 3.   FIND H_L, THE HEIGHT OF THE KEYSTONED IMAGE ON THE LEFT SIDE (OR SHORT SIDE).

NOTE FIRST THAT:

$$\text{TAN } \phi = \frac{\frac{H}{2}}{OP_H} \text{ AND SIN } \theta = \frac{T.D.}{OP_H}$$

$$\text{THEN } OP_H = \frac{T.D.}{\text{SIN } \theta} \text{ AND TAN } \phi = \frac{H\text{SIN}\theta}{2T.D.}$$

$$\text{NOW } \frac{H_L}{2} = \text{TAN } \phi \,(OP_H - W_L)$$

$$= \frac{H\text{SIN}\theta}{2\,T.D.}\,(OP_H - W_L)$$

$$\text{OR } H_L = \frac{2H \text{ SIN } \theta}{2 \, T.D.}\,(OP_H - W_L$$

SUBSTITUTING FOR OP_H:

$$H_L = \frac{H \text{ SIN } \theta}{T.D.}\left(\frac{T.D.}{\text{SIN } \theta - W_L}\right) = H - \frac{HW_L \text{ SIN } \theta}{T.D.}$$

$$\boxed{H_L} = H\left(1 - \frac{W_L \text{ SIN } \theta}{T.D.}\right) \text{ -- -- -- -- -- -- -- -- } (7)$$

STEP 4.   IN A SIMILAR MANNER FIND H_R, THE HEIGHT OF THE KEYSTONED IMAGE ON THE RIGHT SIDE (OR LONG SIDE)½

$$\boxed{H_R} = H\left(1 + \frac{W_R \text{ SIN } \theta}{T.D.}\right) \text{ -- -- -- -- -- -- -- } (8)$$

RECAP

$$\boxed{W_L} = \frac{T.D.(\text{SIN } \beta)}{\text{SIN }(90 - \beta + \theta)} \qquad \boxed{W_R} = \frac{T.D. (\text{SIN } \beta)}{\text{SIN }(90 - \beta - \theta)}$$

$$\boxed{H_L} = H\left(1 - \frac{W_L \text{ SIN } \theta}{T.D.}\right) \qquad H \qquad \boxed{H_R} = H\left(1 + \frac{W_R \text{ SIN } \theta}{T.D.}\right)$$

ST'D IMAGE FORMAT   W

Table 8-1—cont.
VERTICAL Keystone Constants

35-mm 2×2 Slides

ANGLE OF DOWNTILT & ITEM		THROW DISTANCE (IN TERMS OF ST'D IMAGE W)						
		1W	2W	3W	4W	5W	6W	7W
NOTE: ALL VALUES IN THIS TABLE MUST BE MULTIPLIED BY W, THE WIDTH OF THE NORMAL IMAGE.								
6°	H_T	.3280	.3338	.3357	.3367	.3373	.3377	.3380
	H_B	.3523	.3458	.3438	.3427	.3421	.3417	.3414
	W_T	.9657	.9826	.9883	.9912	.9929	.9941	.9950
	W_B	1.0368	1.0181	1.0120	1.0090	1.0072	1.0060	1.0051
7°	H_T	.3268	.3335	.3357	.3369	.3376	.3380	.3384
	H_B	.3552	.3476	.3452	.3439	.3432	.3427	.3424
	W_T	.9602	.9797	.9864	.9897	.9918	.9931	.9941
	W_B	1.0433	1.0212	1.0140	1.0105	1.0084	1.0070	1.0060
8°	H_T	.3257	.3332	.3358	.3372	.3379	.3385	.3389
	H_B	.3582	.3495	.3467	.3453	.3444	.3439	.3435
	W_T	.9547	.9768	.9844	.9883	.9906	.9921	.9933
	W_B	1.0499	1.0243	1.0161	1.0120	1.0096	1.0080	1.0068
9°	H_T	.3247	.3331	.3361	.3375	.3384	.3390	.3395
	H_B	.3615	.3515	.3483	.3467	.3458	.3451	.3447
	W_T	.9492	.9739	.9825	.9868	.9894	.9912	.9924
	W_B	1.0566	1.0275	1.0179	1.0136	1.0108	1.0090	1.0077
10°	H_T	.3238	.3331	.3364	.3380	.3390	.3397	.3402
	H_B	.3648	.3536	.3500	.3482	.3472	.3465	.3460
	W_T	.9438	.9711	.9805	.9853	.9882	.9902	.9916
	W_B	1.0602	1.0307	1.0203	1.0151	1.0121	1.0100	1.0086

Table 8-2.
HORIZONTAL Keystone Constants

35-mm 2×2 Slides

ANGLE OF SKEW & ITEM		THROW DISTANCE (IN TERMS OF ST'D IMAGE W)						
		1W	2W	3W	4W	5W	6W	7W
NOTE: ALL VALUES IN THIS TABLE MUST BE MULTIPLIED BY W, THE WIDTH OF THE NORMAL IMAGE.								
1°	W_L	.4958	.4979	.4986	.4990	.4992	.4993	.4995
	W_R	.5045	.5023	.5015	.5012	.5009	.5008	.5007
	H_L	.6698	.6727	.6737	.6742	.6745	.6747	.6748
	H_R	.6816	.6786	.6776	.6772	.6769	.6767	.6765
2°	W_L	.4917	.4960	.4974	.4981	.4986	.4989	.4991
	W_R	.5092	.5047	.5032	.5025	.5021	.5018	.5016
	H_L	.6641	.6698	.6718	.6727	.6733	.6737	.6740
	H_R	.6817	.6787	.6796	.6786	.6780	.6776	.6774
3°	W_L	.4879	.4942	.4963	.4974	.4981	.4985	.4988
	W_R	.5142	.5073	.5051	.5040	.5033	.5029	.5026
	H_L	.6584	.6669	.6698	.6713	.6721	.6727	.6732
	H_R	.6848	.6846	.6816	.6801	.6792	.6786	.6782
4°	W_L	.4843	.4926	.4954	.4969	.4977	.4983	.4987
	W_R	.5194	.5101	.5071	.5056	.5047	.5042	.5037
	H_L	.6528	.6641	.6679	.6698	.6710	.6718	.6723
	H_R	.6879	.6877	.6836	.6816	.6804	.6796	.6791
5°	W_L	.4809	.4912	.4947	.4965	.4976	.4983	.4988
	W_R	.5249	.5131	.5093	.5075	.5063	.5056	.5051
	H_L	.6474	.6612	.6660	.6684	.6698	.6708	.6715
	H_R	.6911	.6908	.6857	.6831	.6816	.6806	.6799

Table 8-2—cont.
HORIZONTAL Keystone Constants

35-mm 2×2 Slide

ANGLE OF SKEW & ITEM		THROW DISTANCE (IN TERMS OF ST'D IMAGE W)						
		1W	2W	3W	4W	5W	6W	7W
NOTE: ALL VALUES IN THIS TABLE MUST BE MULTIPLIED BY W, THE WIDTH OF THE NORMAL IMAGE.								
6°	W_L	.4777	.4899	.4941	.4962	.4975	.4984	.4990
	W_R	.5307	.5163	.5117	.5095	.5081	.5072	.5066
	H_L	.6419	.6584	.6640	.6669	.6686	.6698	.6706
	H_R	.6944	.6939	.6877	.6847	.6829	.6816	.6808
7°	W_L	.4746	.4888	.4937	.4961	.4976	.4987	.4994
	W_R	.5367	.5197	.5143	.5116	.5100	.5090	.5082
	H_L	.6366	.6556	.6621	.6655	.6675	.6688	.6698
	H_R	.6978	.6971	.6898	.6862	.6841	.6827	.6817
8°	W_L	.4718	.4878	.4934	.4962	.4979	.4991	.4999
	W_R	.5431	.5233	.5170	.5140	.5121	.5109	.5100
	H_L	.6313	.6527	.6602	.6640	.6663	.6679	.6690
	H_R	.7012	.7003	.6919	.6878	.6853	.6837	.6825
9°	W_L	.4691	.4870	.4932	.4964	.4983	.4996	.5006
	W_R	.5498	.5271	.5200	.5165	.5144	.5130	.5120
	H_L	.6261	.6500	.6583	.6625	.6651	.6669	.6681
	H_R	.7047	.7035	.6940	.6893	.6865	.6847	.6834
10°	W_L	.4666	.4863	.4932	.4968	.4989	.5004	.5014
	W_R	.5568	.5311	.5230	.5192	.5168	.5153	.5142
	H_L	.6209	.6471	.6564	.6611	.6640	.6659	.6673
	H_R	.7083	.7068	.6961	.6909	.6878	.6858	.6843

The true keystoned image dimensions are shown in Fig 8-5.

Example 2.

Determine whether or not a dual image 35-mm horizontal slide format is acceptable when the projectors are located at 6 W and the downtilt angle is 8°. The normal image size is 8 ft × 5.4 ft high.

Solution

The chart giving vertical keystone constants shows that the combination of 6 W and 8° is to the right of the heavy line — indicating that the distortion is less than 2%. While this amount of keystone is generally satisfactory for a single image, it may be found unacceptable when side-by-side images are used. It will be necessary to investigate the effect of the wedge of unused screen showing between the images, by drawing to scale the actual configuration.

From the vertical keystone chart we read the following values, each in turn to be multiplied by the normal width of 96 in.

$$H_T = .3385 \times 96 = 32.5''$$
$$H_B = .3439 \times 96 = 33.01''$$
$$W_T = .9921 \times 96 = 95.24''$$
$$W_B = 1.008 \times 96 = 96.77''$$

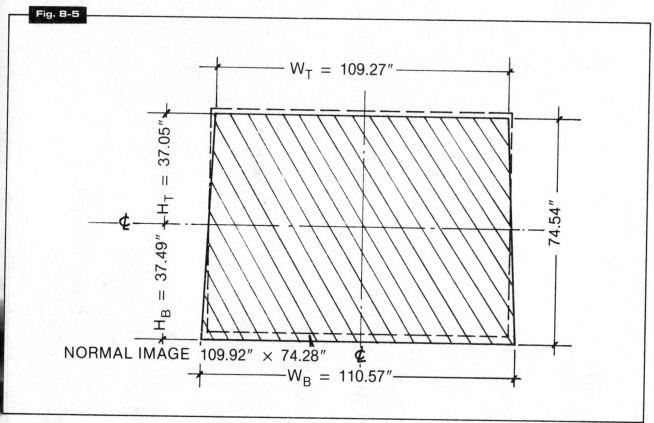

Fig. 8-5

$W_T = 109.27''$

$H_T = 37.05''$

$H_B = 37.49''$

74.54"

NORMAL IMAGE 109.92" × 74.28"

$W_B = 110.57''$

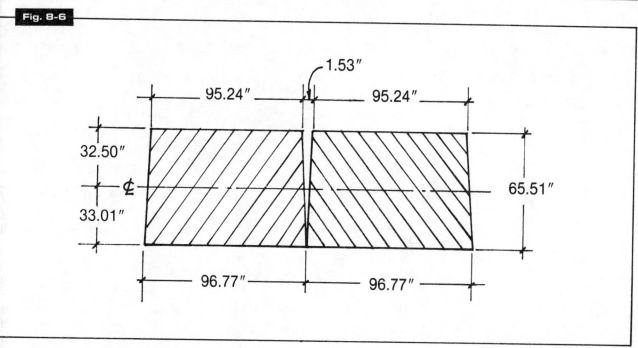

Fig. 8-6

1.53"

95.24"   95.24"

32.50"

33.01"

96.77"   96.77"

65.51"

g. 8-6 is a scale drawing of the two formats as they will ap-
ear on a vertical screen.

It now remains a matter of judgment whether or not the
all triangle between the images, measuring 1½ in wide at
the top, will be acceptable. If it is considered not acceptable,
the keystone can be reduced to a negligible amount by tilting
the screen back 5° at the top. This would leave only 3° of
keystone, with very satisfactory results.

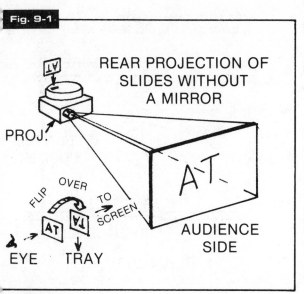

Fig. 9-1

REAR PROJECTION OF
SLIDES WITHOUT
A MIRROR

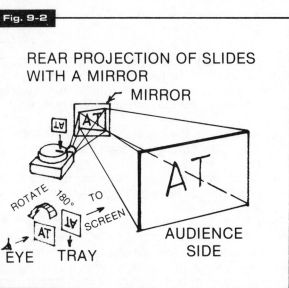

Fig. 9-2

REAR PROJECTION OF SLIDES
WITH A MIRROR

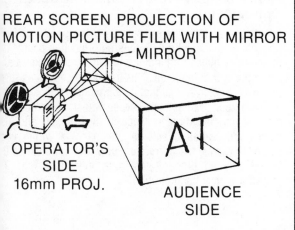

Fig. 9-3

REAR SCREEN PROJECTION OF
MOTION PICTURE FILM WITH MIRROR

SECTION 9

MIRRORS IN PROJECTION SYSTEMS

Mirrors are used in projection systems for two basic reasons: (1) To reverse an image, and (2) To "fold" the light beam. Often there is need to do both. If we take into account all the variables possible when mirrors are used, we find that the possibilities are many. Let us examine several projection situations so that we may become more familiar with the need for the use of mirrors.

1. REAR PROJECTION OF SLIDES WITHOUT A MIRROR (FIG. 9-1)

When slides are rear projected without the use of a mirror, the transparency must be inserted into the projector aperture as shown in illustration number 1. Follow this procedure:

a. Stand behind the projector, looking at the back side of the screen.

b. Raise a slide to the eye and orient it so that it reads correctly.

c. Flip it directly forward, or backward, 180°, so the top edge is now on the bottom, and insert in the slide tray in the desired slot.

The projected image will now be correctly oriented for audience viewing. This projection method has one disadvantage: if the slide tray is used also for front projection elsewhere, all the images will be inverted left-to-right, unless the slides are reoriented, rotating each slide 180° about its vertical axis. The situation can be corrected by loading the tray for front projection, then using a mirror to reverse the image for rear projection, as shown in Fig. 9-2.

2. REAR PROJECTION OF SLIDES WITH A MIRROR (FIG. 9-2)

In this case the projector tray is loaded for front projection by the following method:

a. Stand behind the projector.

b . Raise slide to the eye and orient it so that it reads correctly.

c . Rotate slide 180° to left or right, and insert into tray at desired slot.

The projected image, after left-to-right reversal by the mirror, will now be correctly oriented for audience viewing.

3. REAR PROJECTION OF MOTION PICTURE FILM WITH MIRROR (FIG. 9-3)

When motion picture film is used, the sound track edge of the film must be oriented a certain way to be reproduced by the optical or magnetic sound head. Additionally, in the case of 8-mm and 16-mm film, there is only one row of sprocket holes, and they must be oriented to mate with the sprocketed rollers of the drive mechanism. It therefore is not possible to turn the film around as we do a slide. An image reversing mirror must be used.

The mirrors we have discussed so far are known as "outboard" mirrors because they are generally mounted independently from the projector. In some instances the mirror is rather small, say 6 in × 8 in, but in other cases the mirror may be 4 ft or 5 ft wide. The size depends upon how far along the beam the mirror is placed. The smallest mirror that can be used is placed within a right-angle lens mount, and is part of the lens assembly. It is used when the throw distance needed is within the available depth of the rear screen projection room.

4. REAR PROJECTION WITH LONG LIGHT BEAM BUT LIMITED ROOM DEPTH

When a relatively long focal length lens is used rear screen, such as a 3- or 4-in lens with a 35-mm slide projector, the light beam may be too long for the depth of the projection room. The image reversal mirror now serves to fold the beam as well as to reverse the image. Its size can be considerable. In Fig. 9-4, note that the projector is not placed parallel to the rear wall. If it were, the mirror would have to be mounted at 45° to the wall, and consequently considerably longer in length. The important thing is that the centerline of the beam must be perpendicular to the screen if keystone is to be avoided. The mirror angle may vary, depending on the location of the projector.

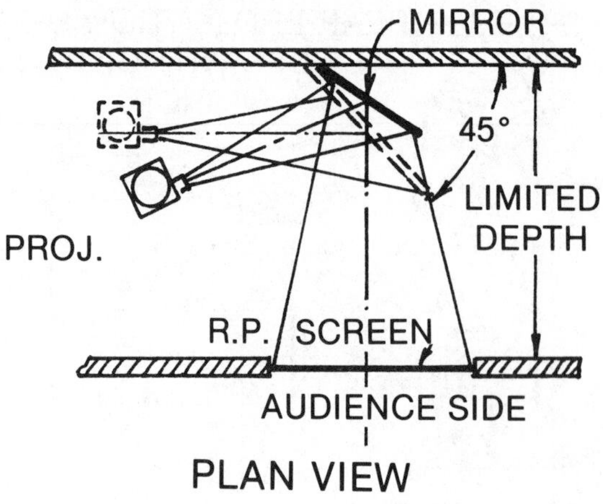

5. REAR PROJECTION WITH TWO MIRRORS TO FOLD THE BEAM, SCHEME 1

In scheme 1 (Fig. 9-5), often used when there is a screen located above a chalkboard, and the projector must be placed at normal floor mounting height, neither the upper nor lower mirror is reversing the image left-to-right. The two mirrors are merely folding the beam without altering the image orientation. This means that we cannot load the tray for standard front projection. If we did, the image would read reversed to the audience. Consequently we must load the tray as described in Fig. 9-1.

Notice in scheme 1 (Fig. 9-5) that the ray emerging from the lens at the top of the beam stays on top all the way to the screen. This is how we know whether or not the image will read correctly from the audience side of the screen.

We will see in scheme 2 (Fig. 9-6) that a ray can start out on the top of the beam and end up on the bottom of the beam!

6. REAR PROJECTION WITH TWO MIRRORS TO FOLD THE BEAM, SCHEME 2

This configuration places the slide projector facing the rear wall instead of the screen wall, as was done in scheme 1. By making this apparently slight rearrange-

Fig. 9-4

REAR PROJECTION WITH LONG
LIGHT BEAM AND LIMITED DEPTH ROOM

MIRROR

45°

LIMITED
DEPTH

PROJ.

R.P. SCREEN

AUDIENCE SIDE

PLAN VIEW

ment, the image is inverted top-to-bottom. The ray th[...] left the projector on the bottom of the image has n[...] appeared at the top of the screen! This makes it nece[...] sary to reorient the slide once more. The procedure is [...] follows:

a. Stand behind the projector looking toward mirror A[...]
b. Raise slide to the eye and orient it so that it reads co[...] rectly.
c. Rotate slide 180° about its vertical axis, so the l[...] side is now on the right. Insert in tray in the desir[...] slot.

There is no particular advantage in using scheme 1 ov[...] scheme 2. It is a matter of preference — perhaps ease [...] operation, or mirror support simplicity. Both schemes cou[...] start off with a mirror-type lens aimed up, if the first leg of t[...] light beam were vertical. This would eliminate mounting pro[...] lems.

7. FRONT SCREEN FILM PROJECTION USING TWO MIRRORS

If a motion picture film projector is being used in a fro[...] projection system where there is no projection booth, t[...] projector is housed oftentimes in a mobile or fixed cab[...] net at the rear of the room. This usually necessitates tur[...] ing the projector 90° so that its long dimension is paral[...] to the screen, rather than perpendicular. This arrang[...] ment saves considerable space in the cabinet, and als[...] places the operating side of the machine within eas[...] reach. However, two mirrors are now required to keep t[...] image oriented properly for front projection. Fig. 9[...] shows a typical configuration.

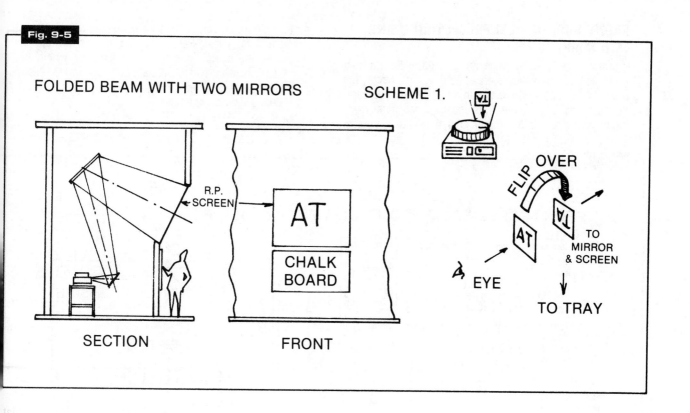

Fig. 9-5

FOLDED BEAM WITH TWO MIRRORS   SCHEME 1.

R.P. SCREEN

AT

CHALK BOARD

SECTION   FRONT

FLIP OVER

AT

EYE

TO MIRROR & SCREEN

TO TRAY

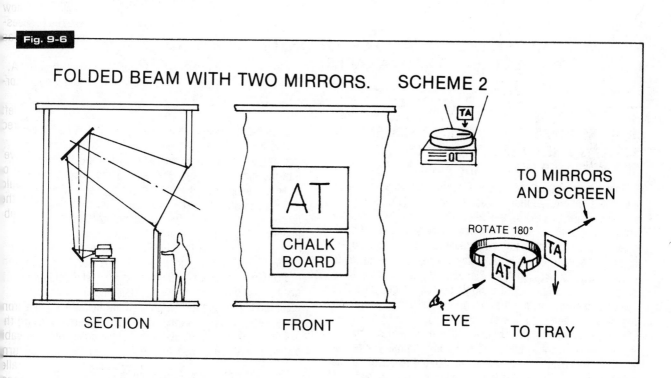

Fig. 9-6

FOLDED BEAM WITH TWO MIRRORS.   SCHEME 2

AT

CHALK BOARD

SECTION   FRONT

TO MIRRORS AND SCREEN

ROTATE 180°

AT

EYE

TO TRAY

7. FRONT SCREEN FILM PROJECTION USING TWO MIRRORS

If a motion picture film projector is being used in a front projection system where there is no projection booth, the projector is housed oftentimes in a mobile, or fixed cabinet at the rear of the room. This usually necessitates turning the projector 90° so that its long dimension is parallel to the screen, rather than perpendicular. This arrangement saves considerable space in the cabinet, and also places the operating side of the machine within easy reach. However, two mirrors are now required to keep the image oriented properly for front projection. Fig. 9-7 shows a typical configuration.

Fig. 9-7

FRONT SCREEN FILM PROJECTION WITH 2-MIRRORS

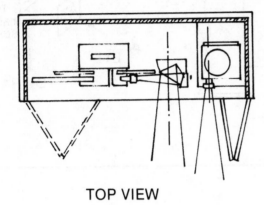

TOP VIEW

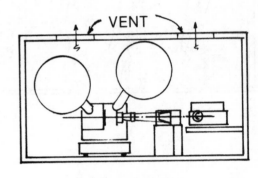

FRONT VIEW

CABINET BUILT-IN SAVES SPACE WHEN THERE IS NO PROJECTION ROOM AVAILABLE.

HOW ARE MIRRORS USED IN REAR SCREEN SYSTEMS WHEN DISSOLVE OR OTHER SUPERIMPOSED IMAGES ARE REQUIRED?

This situation probably causes more problems in image alignment than any other projector configuration. If the designer adheres to a few simple rules, obtaining the desired results is not difficult. When it is necessary to use mirrors for space saving as well as for image reversal, the mirror size becomes rather large. Fig. 9-8A shows a pair of 35-mm slide projectors, nested together horizontally as close as is practical, superimposed on a single screen image. Note that the projection beams are first drawn as though there were no mirror. A convenient scale should be used.

If paper cutouts of the projectors and their light beams are made, the beams may be folded along any suitable line representing the mirror, and the exact location of the mirror can be determined along with that of the projector. Turn the patterns over before folding the beam, then the projector will be right side up after folding. Make sure that no part of the light beam interferes with the projectors or their support stands.

In Fig. 9-8B a television projector having three lenses is shown, one each for the red, blue, and green "guns," all aimed and superimposed on a single converged image on the screen. The use of a mirror across the three light beams will not affect the system. The mirror surface will, however, have to be flat within close tolerance, because here we have three colored beams of light all converging at the screen to produce a sharp image. A sharp image depends upon accurate registration of the three colors. The introduction of a mirror into the system will not alter any of the images even if the three lenses are not in a common line, such as those projecting from the Swiss-made Eidophor color projector. Here the lenses are placed with their nodal points lying in a triangular plane at the front of the unit.

It is worth noting that in Figs. 9-8A and B the mirror is vertical — perpendicular to the floor. This is true even if the projection device is tilted down or up. After folding a light beam, the important requirement is that the projectors are mounted at the identical height and tilt they would have had there were no mirror in the system. If a system works without a mirror, it will work with one, provided proper image orientation is maintained.

Fig. 9-9A shows an elevation view of two 35-mm slide projectors nested vertically, using a space-saving mirror. The plan view is drawn in Fig. 9-9B. First the systems are drawn without the mirror, then the mirror is added. Note that the mirror does not have to be as wide as the mirror of Fig. 9-because the two beams are positioned one above the other. Note also that the mirror is vertical, even though the top projector aims downward, and the bottom projector aims upward, both images converging at the screen.

As mentioned earlier, projectors must not be tilted toward

their side, only toward their front or back, unless mirror type lenses are being used. The same up-tilt and down-tilt as required in the nonmirror situation must be maintained.

If dual image dissolve is required with either arrangement of Fig. 9-8 or 9-9, the configuration is simply repeated, with the projectors placed to the right instead of to the left of the mirror. The mirrors will form a broken "V" with their optical centers spaced apart a distance equal to the image width.

In comparing layouts shown in Figs. 9-8 and 9-9, each has its own particular advantage. Although the mirror is more costly in Fig. 9-8, due to its large width, it has the advantage of accommodating high-intensity light source projectors without increasing the lens-to-lens distance, thus aggravating the keystone condition. In Fig. 9-9, however, if the projectors are increased in height to accommodate the blower section needed for the higher wattage lamps, the spread between lenses becomes unacceptable, increased even further by the need to leave space for tray removal on the lower projector.

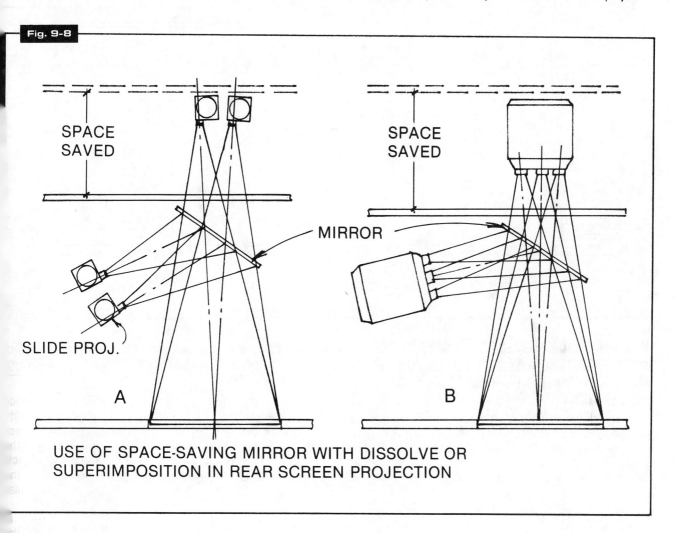

Fig. 9-8

SPACE SAVED

SPACE SAVED

MIRROR

SLIDE PROJ.

A

B

USE OF SPACE-SAVING MIRROR WITH DISSOLVE OR SUPERIMPOSITION IN REAR SCREEN PROJECTION

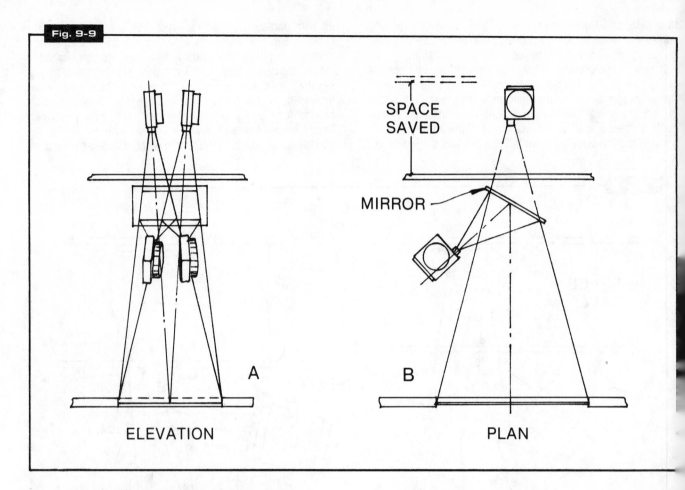

Fig. 9-9

SPACE
SAVED

MIRROR

A

ELEVATION

B

PLAN

IN 35-mm SLIDE PROJECTION, WHAT IS THE MOST PRACTICAL WAY TO PROVIDE FOR THE VARIABLES THAT HAVE BEEN DISCUSSED ABOVE?

Unless there is a definite need for space saving beyond that required by the normal use of a 2-in focal length lens in a rear projection system, the arrangement shown in Fig. 9-10 is probably the most practical. It makes use of new, high resolution, large aperture mirror lenses on 35-mm slide projectors of the carousel type, superimposes dissolves with only 2° of horizontal keystone, and accommodates all types of high-intensity light sources.

Additionally, it allows simple mounting of projectors on a single flat table or shelf. Slides are inserted for front projection so the trays are interchangeable with front projection systems. Distortion of the image due to keystone is approximately 2%, which is generally satisfactory.

Fig. 9-10

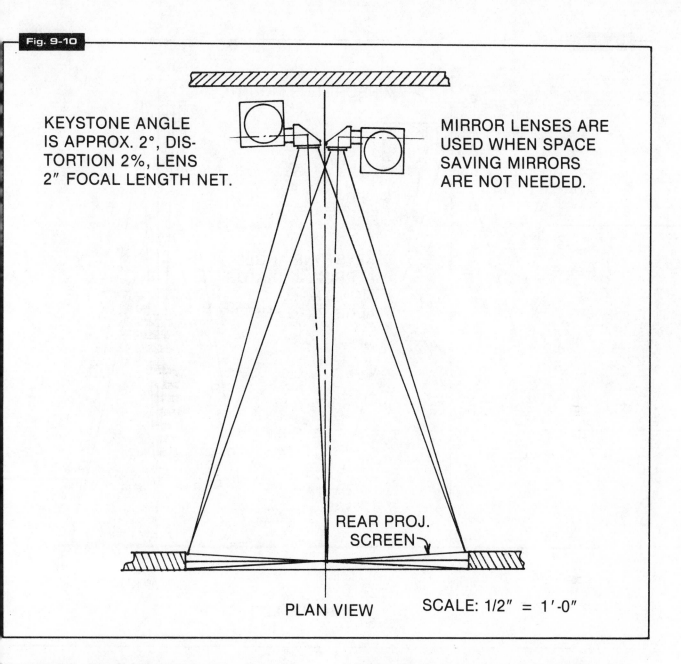

KEYSTONE ANGLE IS APPROX. 2°, DISTORTION 2%, LENS 2″ FOCAL LENGTH NET.

MIRROR LENSES ARE USED WHEN SPACE SAVING MIRRORS ARE NOT NEEDED.

REAR PROJ. SCREEN

PLAN VIEW   SCALE: 1/2″ = 1′-0″

IS VERTICAL PROJECTOR STACKING PRACTICAL, USING MIRROR-TYPE LENSES?

In order to provide the advantage of being able to use high-intensity light sources, keep keystone to a minimum, and permit tray removal, it is recommended that vertically stacked projectors be stacked with only their lens mounts in line, one above the other. The arrangement is shown in Fig. 9-11. Keystone is kept down to 2° with this design.

Note that the entire projector and its lens mount is aimed up or down, as the case may be, at an angle just large enough to avoid interference, one with the other. The only disadvantage with this arrangement is the need for a rather complicated projector support framework.

IS THERE A QUICK, RELIABLE FORMULA FOR PREDICTING THE REQUIRED DEPTH OF THE PROJECTION ROOM FOR A REAR PROJECTION

SYSTEM WHEN SPACE SAVING MIRRORS ARE NOT REQUIRED?

Yes. This formula is valid for rear screen systems using a 2-in net focal length lens, with or without a mirror lens attachment.

$$\text{Clear depth of projection room} = \left[\frac{\text{Distance from screen to farthest viewer, ft}}{3.6}\right] + 2 \text{ ft}$$

Example

Find the minimum depth of a rear projection room for a lecture hall, where the distance from the screen to the farthest viewer is 40 ft.

Solution

$$\text{Depth} = \left[\frac{40}{3.6}\right] + 2 = 13 \text{ ft-0 in}$$

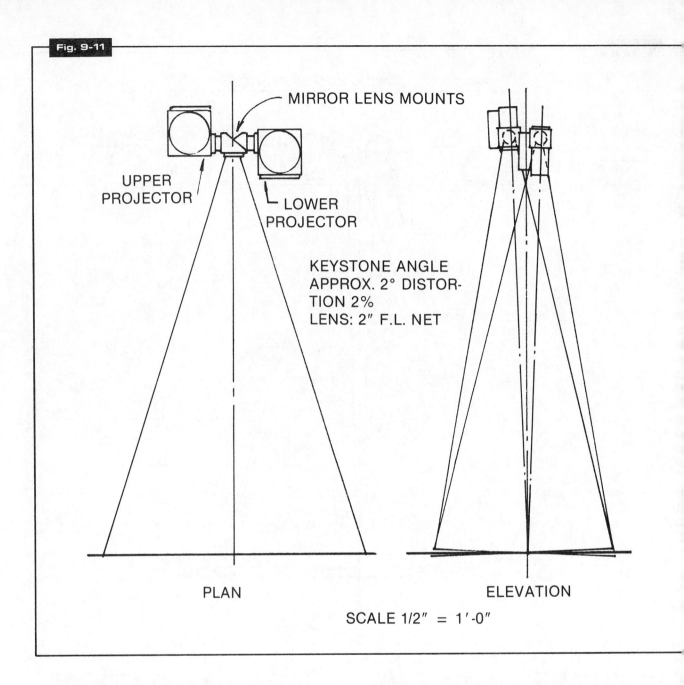

Fig. 9-11

MIRROR LENS MOUNTS

UPPER
PROJECTOR

LOWER
PROJECTOR

KEYSTONE ANGLE
APPROX. 2° DISTOR-
TION 2%
LENS: 2" F.L. NET

PLAN

ELEVATION

SCALE 1/2" = 1'-0"

WHAT KIND OF GLASS MIRRORS ARE USED IN PROJECTION APPLICATIONS, AND HOW ARE THEY MOUNTED?

Mirrors for this kind of optical use are made of high quality float glass, free from irregularities, striae, and nonuniform coloration due to impurities. Thickness is usually ¼ in for sizes up to 4 ft wide, and ⅜ in for larger sizes.

The reflective layer is vacuum coated, and is usually aluminum fluoride. The finish is a high polish, 90% reflective surface on the front face of the glass. Such mirrors are known as front, or first surface, mirrors, in contrast to the more familiar decorative mirrors which are silvered on the back side, and called second surface mirrors.

The front surface coating eliminates ghost images caused by double reflection from both front and rear surfaces of common mirrors, but it creates a serious problem of vulnerability to scratches and atmospheric oxidation. Front surface mirrors are usually coated with a thin clear epoxy to increase their resistance to corrosion.

Great care must be exercised in mounting a mirror so that the glass is never stressed by the mounting lugs, causing it to deflect or warp. It takes only a 0.005-in deflection to affect the image. This immediately alerts the installer not to mount the mirrors on plywood backing!

Warpage of the backing panel will distort the glass. Practically, the safest mount is made of welded or bolted steel or aluminum angle, braced for rigidity. The glass is lightly clamped to neoprene pads by plastic clamps, and allowed to seek its own level.

In some designs, the frame is hinged to the wall along one edge, and the angle of tilt easily adjusted. "Tricky" designs, which attempt to use a pivoting mirror on a vertical pole, should be avoided — especially those that are motor driven to a detented position. Before any mirror mount is fastened in place, its alignment should be checked. Once a mirror is out

98

of alignment, the final image cannot be corrected by fudging the projector mounts, or skewing the projectors.

The main reason that installations with complicated mirror mounts cause so much trouble is due to the difficulty of measuring distances and angles in space. Most installers depend on a flexible tape measure, a 3-ft level, and their eye to ensure proper alignment of optical components. A transit and stadia-rod target would be far more accurate, and a lot less time consuming.

WHAT IS MEANT BY SECONDARY IMAGE, AND HOW DOES IT DIFFER FROM GHOST IMAGE?

A ghost image is the result of a closely spaced double reflection caused by the thickness of a second surface mirror. It is only slightly displaced from the main image, but results in a lack of sharpness around the edge of the features.

The secondary image is altogether different. It can happen with either a front or second surface mirror. Recall that the audience side of a glass rear projection screen is the coated side, and the back side is regular polished glass. When a large space saving mirror is used to fold the beam, it is possible for light rays at certain angles to reflect off the back side of the screen, strike the mirror, and reflect once more onto the screen. This creates the secondary image. It is theoretically possible for this process to continue, each time leaving a larger but weaker image on the screen. Practically, this doesn't happen, and the secondary image dies after one repeat reflection. The third reflection path usually misses the screen altogether, and is lost in the room.

The ray diagram in Fig. 9-12 explains how the secondary image is formed, and the front view of the screen shows what it looks like. Secondary images are disturbing when a slide contains an opaque background with a brilliant caption. Note that the secondary image has traveled further than the primary image, and hence is proportionately larger. It is also noticeably dimmer, due to the loss of light with each reflection.

It seems fairly obvious that some point on the screen, such as B, will fulfill the conditions needed for secondary reflection. The reflection will not be an entire word, if we are talking about a caption, but only as much of the image as falls within the group of rays near point B will appear as a secondary image.

HOW CAN SECONDARY REFLECTION BE AVOIDED?

Fig. 9-13 explains the mathematical process that evaluates the conditions for secondary reflection. The same analysis can be used to tell us how to avoid secondary reflection. If we examine the figure carefully, we discover that all points to the left of point B either do not strike the mirror at all, or if they do, they are reflected onto the wall, missing the screen altogether.

Point B becomes the critical point, as its reflection falls just at the end of the mirror, point E. If we rotate the mirror clockwise about E, we can cause ray EF to also rotate clockwise about E, so that point F falls just off the left end of the screen. In such a position of the mirror, the reflection of all points to the left of B which now have secondary reflection will fall outside the screen, and all points to the right of B will miss the mirror completely.

For every distance M where we might want to position the mirror, there can be found an angle α that will ensure that the second reflection does not fall on the screen. The minimum tilt of the mirror, α, measured from a line through E at distance M from the screen can be found by the method outlined in Fig. 9-14.

Fig. 9-12

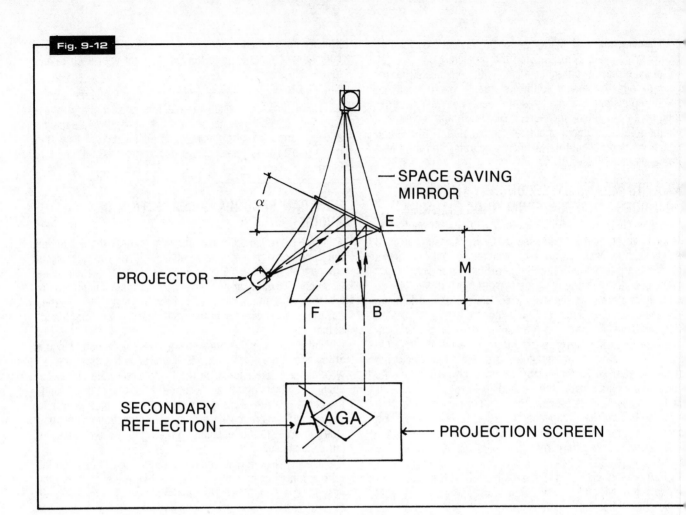

SPACE SAVING MIRROR

E

M

PROJECTOR →

α

F   B

SECONDARY REFLECTION → AGA ← PROJECTION SCREEN

Fig. 9-13

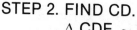

HOW TO FIND FB IN TERMS OF W,M, AND T, WHERE B IS THE CRITICAL POINT THAT COULD REFLECT A RAY TO THE MIRROR AT POINT E, AND THEN TO THE SCREEN ALONG EF. α IS THE LIMITING MIRROR ANGLE. IF α IS REDUCED, POINT F WILL FALL ON THE SCREEN.

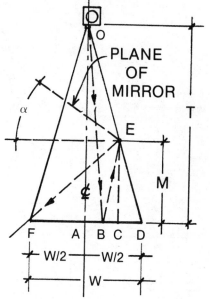

STEP 1. DRAW LIGHT BEAM TO SCALE. LOCATE DESIRED MIRROR POSITION E. DRAW EC ⊥ TO SCREEN.

STEP 2. FIND CD.

$\triangle$ CDE ~ $\triangle$ ADO

$\therefore \dfrac{CD}{\frac{W}{2}} = \dfrac{M}{T}$ AND CD $= \dfrac{WM}{2T}$

STEP 3. FIND BC.

$\triangle$OAB ~ $\triangle$ EBC

$\therefore \dfrac{BC}{AB} = \dfrac{M}{T}$ AND BC $= \dfrac{(M)AB}{T}$

STEP 4. FIND AB IN TERMS OF BC, CD, & W/2

AB + BC $= \dfrac{W}{2} -$ CD. NOW SUBSTITUTE 2 & 3 IN 4:

$AB = \dfrac{W(T - M)}{2(T + M)}$

STEP 5. FIND FB: FB $=$ FA + AB $= \dfrac{W}{2} + \dfrac{W}{2}\left(\dfrac{T - M}{T + M}\right)$

OR FB $= \dfrac{W}{2}\left[1 + \dfrac{T - M}{T + M}\right]$

Fig. 9-14

GRAPHICAL PROCEDURE TO CHECK FOR SECONDARY
REFLECTION IN MIRROR SYSTEM

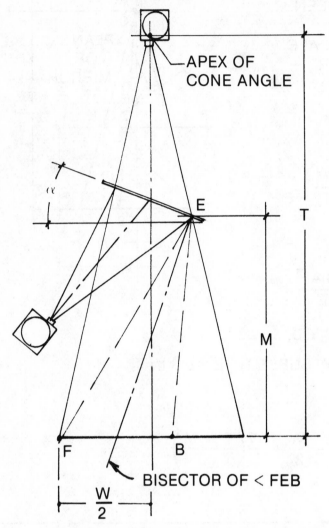

APEX OF
CONE ANGLE

BISECTOR OF < FEB

$\frac{W}{2}$

STEP 1. CALCULATE:

$$FB = \frac{W}{2}\left[1 + \frac{T-M}{T+M}\right]$$

AND LOCATE B TO
SCALE OF DRAWING.

STEP 2. DRAW FE & BE.

STEP 3. BISECT ANGLE
FEB.

STEP 4. DRAW MIRROR
PERPENDICULAR
TO BISECTOR.

STEP 5. MEASURE ANGLE
α WITH PROTRAC-
TOR. α IS A MIN-
IMUM FOR MIRROR
LOCATION AT E.

STEP 6. "FOLD OVER" LIGHT
BEAM AND LOCATE
PROJECTOR LENS.

NOTE: IF PROJECTOR BODY INTERFERES WITH LIGHT
BEAM, INCREASE α TO SUIT. THERE WILL BE NO SEC-
ONDARY REFLECTION FOR MIRROR ANGLES EQUAL TO
α OR GREATER.

MIRROR TOUCHING SCREEN
AT M = 0

α SHOWN IS MINIMUM.
α SHOULD BE INCREASED
TO ALLOW CLEARANCE BE-
TWEEN PROJECTOR AND
SCREEN WALL.

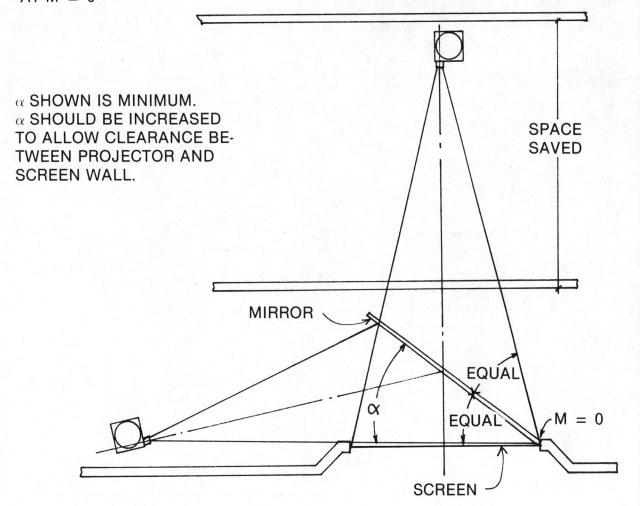

SPACE SAVED

MIRROR

EQUAL

EQUAL

M = 0

α

SCREEN

THIS ARRANGEMENT IS A REAL SPACE SAVER, BUT RE-
QUIRES A VERY LARGE MIRROR, HENCE NOT TOO PRAC-
TICAL FOR LARGE IMAGE SIZES. FINDS USE IN OPTICAL
EQUIPMENT WHERE MIRROR SIZE IS SMALL.

We now have a simple way of finding the critical point B on the screen, by measuring to scale the distance FB found in step 5 of the mathematical proof. The mirror angle now becomes the result of a graphical procedure using the distance FB, as outlined in Fig. 9-14.

CAN ONE END OF A SPACE SAVING MIRROR ACTUALLY TOUCH THE SCREEN (M = 0)?

Yes. The angular position of the mirror is found in this case by placing it on the line which bisects the corner angle formed by the screen and the edge ray. This case is illustrated in Fig. 9-15.

HOW IS THE SIZE OF THE MIRROR DETERMINED?

The size of the mirror is not difficult to find, but it will be necessary to draw the light beam and mirror location to scale. Fig. 9-16 explains the procedure in a step-by-step method.

Fig. 9-16

DETERMINING MIRROR SIZE

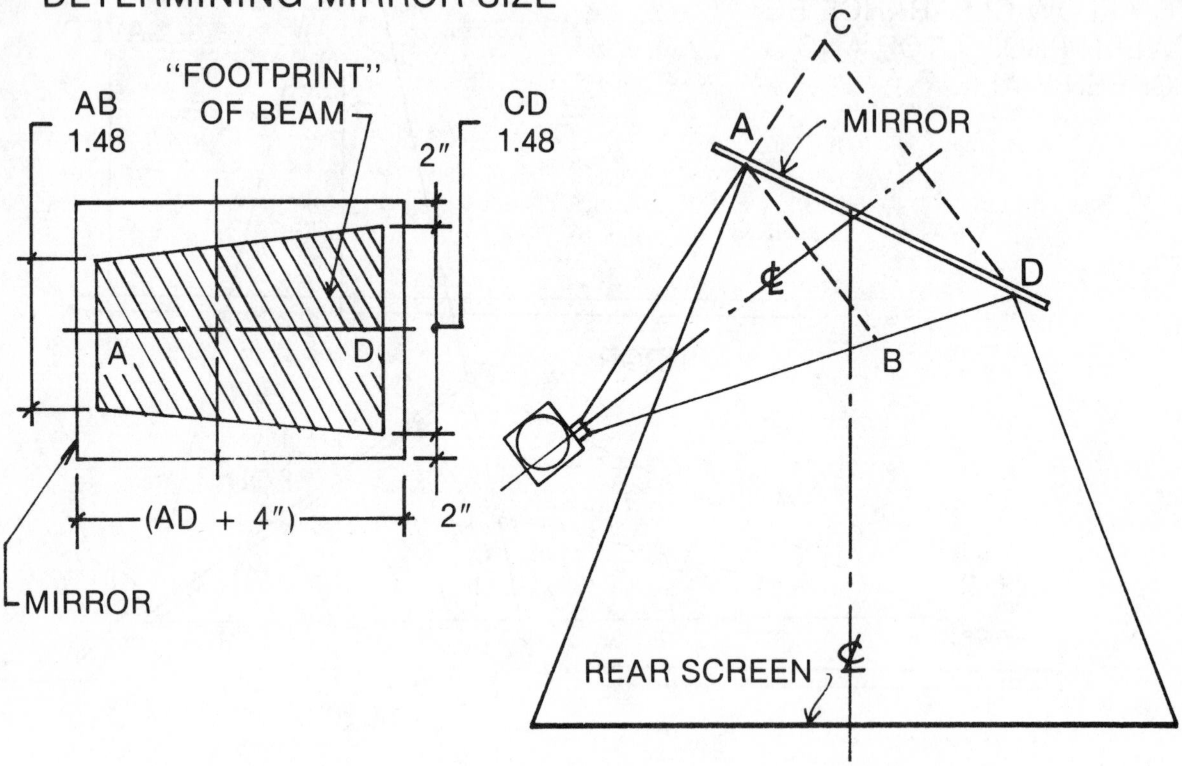

STEP 1. DRAW AB AND DC PERPENDICULAR TO OPTICAL ₵

STEP 2. MEASURE THE IMAGINARY IMAGE WIDTH AB TO SCALE
OF DRAWING. IMAGE HEIGHT @ A IS AB ÷ 1.48
1.48 IS ASPECT RATIO OF 35−MM SLIDE.

STEP 3. REPEAT FOR IMAGE HEIGHT @ D = CD ÷ 1.48

MIRROR WIDTH = AD + 4″ (SAFETY MARGIN)
MIRROR HEIGHT = STEP 3 + 4″
MIRROR IS USUALLY MADE RECTANGULAR, ALTHOUGH
IT COULD BE CUT TO TRAPEZOIDAL SHAPE.

THE VIDEO IN "AUDIO-VISUAL"

The material discussed in this section brings into focus the increasing role that television is playing in almost every kind of audio-visual facility being designed today. Its aim is to discuss television from the user standpoint. How does the planner provide for its use? How is it applied to a given situation? How does its use augment conventional audio-visual projection and display techniques? How does it compare from the standpoint of image quality, audio quality, resolution, image size, system integration, programming preparation, cost, and so forth?

Like any other medium, the popularity of television in communications is directly proportional to the support given to it by the television and related industries — to produce quality equipment, to provide standards of operation, to proliferate in a competitive market, and to make available the accessories, services, and techniques that constitute today's television needs. In these respects, the television industry has not been lacking. As a matter of comparison, it has made greater strides in much less time, and at a higher technical level, than has been evident in the audio-visual industry. Likewise, computer technology has demonstrated still a faster rate of growth.

This situation is normal, for our technical growth follows the curve of learning — ever increasing! Such rapid advancement of technology in the communication field does not mean that video tape will replace film, any more than projection devices replaced the chalkboard, or television replaced the radio. It does show, however, that each new technology will impinge on all previous relevant technologies, elevating each to new standards of excellence. Thus, computers are designing superb optics for all our lens needs, microprocessors are opening up exciting control capabilities for all programmable requirements, and digital video techniques will soon provide us with an all-new concept of television signal generation and transmission.

WHAT WERE THE CIRCUMSTANCES THAT BROUGHT TELEVISION INTO THE AUDIO-VISUAL SCENE?

Educational institutions and corporations with training facilities were among the first group communicators to expand their audio-visual facilities to take advantage of the televised image, to teach and to train. In retrospect, although less than 20 years ago, these users were pioneers. They spent hundreds of thousands of dollars attempting to ride the wave of what was going to be the future way to educate and to train. In this respect, the educational community waded into the deepest water. It was motivated by the excitement of much activity in the exploration of the learning process. Reminders of such activity are the endless human behavior articles that appeared in the many communications and psychology magazines; the introduction of programmed self-instruction machines; the learning laboratories; the study carrel; student response systems; and more recently, computer-assisted instruction.

In this early period of development, it is interesting to note that it was usually some educational learning laboratory that established the criteria for the "ultimate" system. And such systems were often so advanced that many of the needed devices were not commercially available, and had to be developed by some reliable manufacturer willing to risk development costs to establish a possible future proprietary system. A classic example of such an alliance in the early '60s took place when a wealthy school system in a Chicago suburb introduced what was probably, at that time, the most costly and sophisticated learning system in existence. It featured individual study carrels, immediate dial-up random access retrieval of dozens of television tape programs displayed on individual monitors, with earphone sound. But outside of a handful of competitive systems — the extremely expensive random access retrieval, video-tape television system for individual instruction did not become the wave of the future. Few boards of education could justify the cost, and the mere passage of time was producing changes, improvements, and new ideas in competitive techniques.

During the same period, industry was quick to recognize the advantage of the televised image in the areas of classroom training, on-the-job training, role playing, and display of live televised pick-up of key corporate officers addressing the employees via closed-circuit television. Such activities could profit quite naturally from the use of an in-house television studio. The late '60s and early '70s saw the rapid rise of the in-house studio, growing from an 18-ft × 25-ft monaural facility to a 40-ft × 60-ft and larger, double story, fully equipped color studio.

Out of all this flurry of television activity, one thing remained clear — televised teaching, training, and group communication were here to stay. Its value was established not only as a medium, but also as a means to an end. It came at a time when public school enrollment doubled in the 20-year period prior to 1970. Teacher shortage was threatening the educational system. But by 1961 approximately 7500 elementary and secondary schools in 600 school districts across the country were offering at least part of their regular curriculum via television. Intra-school tv exchanges among school systems were taking place with mutual benefit to all parties. Video tapes, kinescope recordings, and live transmission were all used to provide the optimum mix of instructional communication.

By 1962, over 6 million students were benefiting from televised instruction, with half that number attending classes using ITV, and the remainder being reached by ETV. The terms ITV (Instructional Television) and ETV (Educational Television) are sometimes used interchangeably, but there is a basic difference. The acronym ITV refers to formalized educational broadcasts that are transmitted on a planned schedule, from which formal examinations are taken and scholastic credits are earned. If the primary purpose of the material being telecast is to enhance the viewer's knowledge and general understanding of a subject, without giving credits that may be applied to earning a diploma or degree, the term "educational television" applies. The former is directed to a specific viewing group, while the latter ETV is informal. Both make excellent use of closed circuit television techniques.

Many schools make use of a "hub" school which acts as an origination center for microwave transmission to satellite schools. A frequency of 2500 megahertz has been assigned to these line-of-sight systems, and they are referred to as ITFS (Instructional Television Fixed Service).

We see that there is considerable flexibility in the way in which television signals can be delivered to a wide spectrum of viewers. Three kinds of video distribution systems are shown schematically in Figs. 10-1 through 10-3.

ELEMENTARY BASEBAND VIDEO DISTRIBUTION (With "Looping" at a Single Location)

An elementary video baseband distribution system makes use of a single coaxial cable which distributes the video voltage output produced by a television camera, or by a video tape player, directly to television monitors. When there are two or more input sources the signal may be routed through a switcher, so that selecting the desired input is accomplished by pushbutton operation. Note that the audio signal is carried on a separate audio line.

For two or three monitors in the same area, "looping" the signal from one set to the next is possible, but not advisable. With this arrangement, the signal will deteriorate from monitor to monitor, depending on the length of the cables. If one monitor malfunctions, it may affect the operation of the others. Fig. 10-1 shows this rather limited arrangement. This system may be properly called a closed-circuit television system, abbreviated CCTV. The system is "secure."

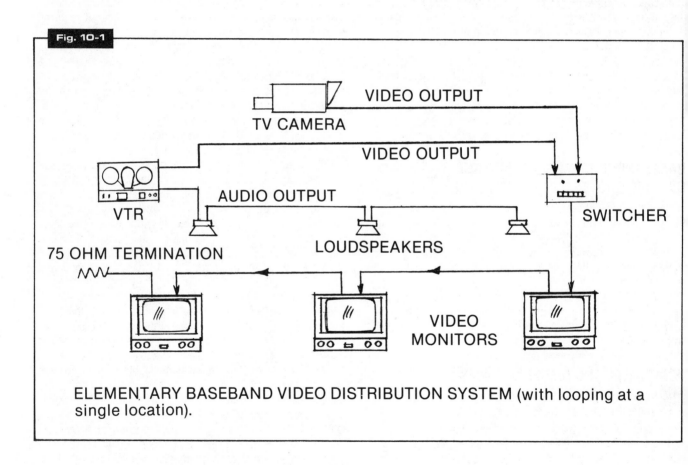

Fig. 10-1

ELEMENTARY BASEBAND VIDEO DISTRIBUTION SYSTEM (with looping at a single location).

BASEBAND VIDEO DISTRIBUTION (Using One or More Video Distribution Amplifiers)

Fig. 10-2 shows a baseband system capable of serving many locations within a given building. It makes use of one or more video distribution amplifiers, which are available with up to six outputs. Audio distribution amplifiers are likewise used to distribute the audio signal, which is usually fed to a room loudspeaker system. When the television system is part of the regular audio-visual system, they share the same audio system. Although Fig. 10-2 shows two monitors looped together on each feed cable, it is recommended that each monitor have a separate home run coaxial cable connected to the central control area.

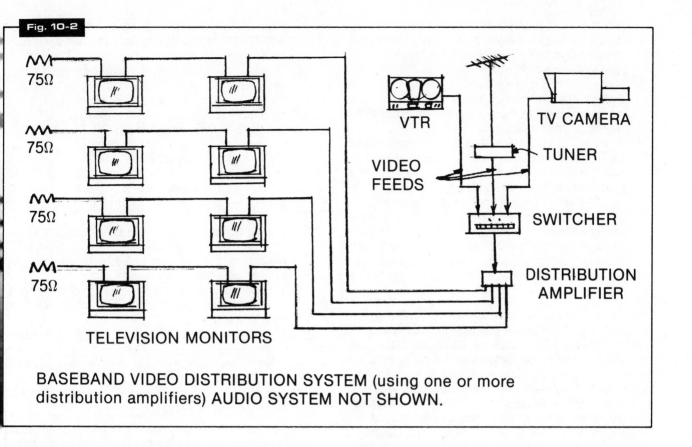

BASEBAND VIDEO DISTRIBUTION SYSTEM (using one or more distribution amplifiers) AUDIO SYSTEM NOT SHOWN.

BASEBAND SYSTEM FEATURES AND CONSIDERATIONS

Here are some important features and considerations when designing a baseband system.

1. Images transmitted via baseband distribution are capable of generally higher quality than those produced by the popular rf systems, to be discussed later. The dedicated baseband system can accommodate high resolution (''hi-rez'') camera and monitor equipment useful in certain laboratory and medical systems. The bandwidth of the transmitted signal is restricted only by the type of cable and length of run, and amount of equalization provided.

2. An inherent system ''plus'' is the capability of being able to plug in a camera or tape player/recorder, and an audio signal source at the receiving end of the system and transmit to the head end or central control location. Thus the video and audio cables are bidirectional.

3. The system is private and secure.

4. Selection of signals is made at the central distribution point, by using a routing or crossbar switcher, or by means of a patch panel. Inasmuch as there is no tuner in the monitor, the viewer has no control over the signal being transmitted other than to adjust it and turn it on and off.

5. Cable cost is high due to the requirement of individual home runs. In a high-rise building, hundreds of cables might be routed vertically, requiring a cable shaft and appropriate support for the dead weight of the cables.

6. Monitors are generally more expensive than television receivers, but they provide higher quality in the image.

Combination monitor/receivers will be discussed later.

7. When more than one program is required for distribution to different areas within a given building, a separate source is required for each program. The routing of any given program is, or course, under the control of the rf distribution system.

RF (RADIO FREQUENCY) DISTRIBUTION SYSTEM

The rf distribution system, sometimes called *modulated* or *broadband* system, makes use of a high frequency carrier wave that is modulated, or shaped, by the video signal. Fig. 10-3 depicts a typical composite of the various distribution capabilities of such a system.

Note that the devices whose outputs are video voltages, such as video cassette recorders and video cameras, must have their output signals modulated at the carrier frequency of an unused channel, so that they appear as rf signals at the *combiner*, where they are combined with the local off-air channels received by the master antenna located on the roof of the building. All the rf signals can now be distributed on a single coaxial cable to the various user devices. Where cable lengths are long, and many sets are served, line amplifiers are required to make up for the losses due to cable resistance and device insertion losses. Here are some important features of the rf system and its components.

1. Both audio and video signals may be fed to the channel modulators. The output of the modulator is in every way like the rf signal received from the broadcast transmitter via the house antenna.

2. After the video signals have been modulated, and mixed with off-the-air signals, they are distributed on a single

coaxial cable. This type of distribution is similar to what is known as cable television. While cable systems serve an entire community or city, the same techniques are used in a single building system.

3. Devices called directional couplers provide a means of drawing off part of the signal at desired locations along the coaxial main, thus feeding one or more television receivers, or other devices, with an rf signal.

4. A splitter, which functions in a manner opposite from the combiner, is used to divide the incoming signal to feed two or more outputs. The two-output splitter is used most, as its signal loss is only 3.5 dB.

5. Fig. 10-3 shows three splitters: one feeds an fm radio and a television receiver; the second, used in conjunction with a switch, provides a convenient way to maintain a permanent connection for taping off-the-air programs or viewing them on a receiver, or both, and playing them back. The third splitter simply feeds two receivers.

Note that in the case of the video cassette recorder shown, its output is rf, inasmuch as it is feeding a receiver, not a monitor. Generally VCRs contain two switchable outputs, one for rf and the other for baseband video. For professional use the output is usually video, because of the high-quality video signal. For home or office use, tv sets are usually receivers, not monitors. Hence the need for the rf output on the cassette machine. Newer television sets, both here and abroad, are circuited for both kinds of signals. Selection is made by a switch on the unit. This type of set is called a receiver/monitor.

WHAT IS THE DIFFERENCE BETWEEN A TELEVISION RECEIVER AND A TELEVISION MONITOR?

The television monitor is designed to display the television signal appearing as a video voltage on a coaxial cable, originating at the video camera, video tape player, or other analog device. Its features and characteristics are:

1. It is not tunable. That is, there is no tuner, or channel selector, because the monitor receives only the analog voltage coming from the source. Consequently it cannot be used as a channel-selecting television receiver.

2. It is a display device only, and does not contain an audio amplifier and loudspeaker system. Program sound must be transmitted separately.

3. Its image quality is excellent, because its signal is relatively pure, not having to undergo modulation, broadcasting, and demodulation at the receiver. Its signal is not subject to ghosting caused by random interference of building reflections and so forth. It is capable of accommodating high-resolution systems.

4. The monitor must be cable-connected to the source. It therefore constitutes a secure link. Each monitor requires a discrete source if it is to display a different program. Line amplifiers are required if cable runs are lengthy.

5. If a monitor is required to display a broadcast signal, that signal must be received on an antenna, tuned to the desired channel, then distributed on the discrete cable system connected to that monitor.

6. The operating controls on a monitor can control on/off, brightness, contrast, horizontal and vertical hold, and color adjustment.

7. High quality monitors are capable of 800 lines of horizontal resolution.

8. Monitors are available in tube diameters ranging from 5 in to 25 in.

9. Monitors are relatively expensive, costing up to four times the cost of a standard receiver.

The television receiver is capable of receiving off-the-air signals, tunable over the full range of uhf and vhf channels, as well as the lettered channels. The tv receiver has the following features:

1. It contains a tuner, so that different channels can be selected.

2. It contains an audio amplifier and loudspeaker system, so that it can reproduce the sound that accompanies the rf television signal.

3. Its image quality can, under favorable conditions, approach that of the monitor.

4. It is capable of 320 lines of horizontal resolution, and 340 lines of vertical resolution in the black and white format, less when transmitting color.

Let us now see how the receiver, or monitor, or receiver/monitor is properly used in AV systems.

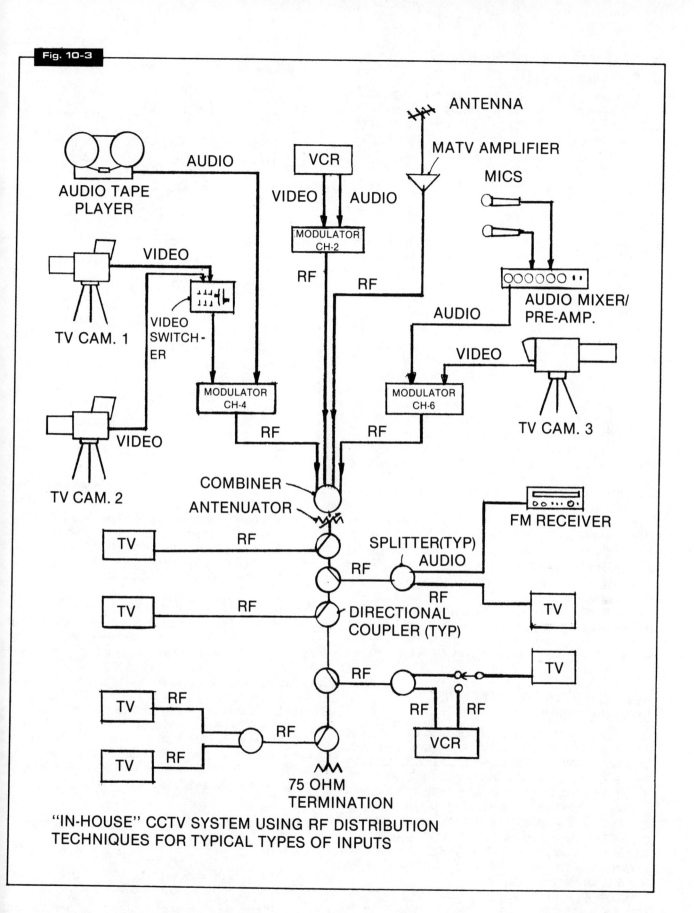

Fig. 10-3

"IN-HOUSE" CCTV SYSTEM USING RF DISTRIBUTION
TECHNIQUES FOR TYPICAL TYPES OF INPUTS

WHAT IS MEANT BY RESOLUTION IN A TELEVISION IMAGE?

The term resolution is used more often than it is understood. It is expressed in lines, and therefore easily confused with the term *scan lines*. It is quite common knowledge that our American television standard is based on the use of an electron beam scanning rate of 525 lines per image frame. These scan lines, which "paint" the image on the picture tube, are tilted slightly downward, starting from the center of the upper edge of the image, and proceeding toward the center of the lower edge of the tube. The actual scanning process takes place in two top-to-bottom scans of 262½ lines each. Each scan is called a field, and there are two fields per frame. The lines of the second field fall exactly between the lines of the first field. These scanned fields occur at an extremely rapid rate, producing 30 frames per second, or 60 fields per second.

In all, there are 15,750 lines per second scanning the television raster area of the tube face. Even though this information is traced on the tube face a line at a time, the persistence of vision of the human eye, combined with the retention qualities of the tube coating phosphors, act to produce a non-flickering photographic image on the brain.

Resolution lines do not refer to an actual number of lines on the screen, but to a numerical standard against which the image detail can be measured. The numerical standard is in the form of a printed black and white chart called a *test pattern*. It contains, among other test configurations, a set of horizontal and vertical wedges consisting of alternate black lines and white spaces. Fig. 10-4 shows how the vertical and horizontal wedge lines are printed on the chart.

In studio use, the cameraman focuses his camera on the chart, and adjusts its optical system for optimum sharpness by visually following each wedge seen on the viewing monitor, toward its tapered end, and reading the number at which the eye can no longer distinguish, or resolve, the separation between the lines.

If the lines on the vertical wedge appear to merge at 320, then we are measuring a horizontal resolution of 320 lines. Likewise, if we find that the lines on the horizontal wedge appear to converge at 340, we are measuring a vertical resolution of 340 lines.

HOW DO WE INTERPRET THE LINES OF RESOLUTION?

For resolution to have any real meaning, we have to do more than start with good resolution at the camera. The entire television system — studio cameras, tape machines, transmitters, antennas, home receivers, and finally the viewer himself are all involved. The viewer is involved because one may not be able to use all the resolution the system can produce due to limitations of the human eye. Obviously the viewing distance, in comparison to the size of the tube being viewed, is an important factor. Before examining what the eye can see, let us consider what is happening on the face of the television tube.

Assume we are looking at a 21-in diagonal picture tube screen size showing 350 lines of vertical resolution on the raster, as read from the transmitted image of a test chart. This means that 350 line-pairs (a black bar and its adjacent white space) would just fill the height of the screen. The

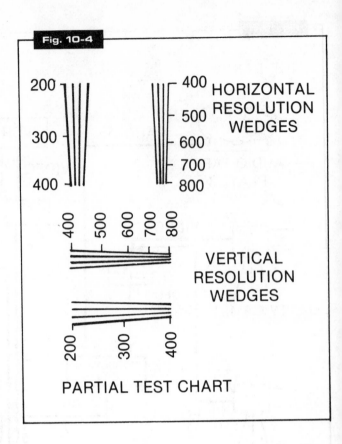

Fig. 10-4

HORIZONTAL RESOLUTION WEDGES

VERTICAL RESOLUTION WEDGES

PARTIAL TEST CHART

height of a 21-in diagonal image is 13¾ in, therefore each line-pair spans a height of 13.75 ÷ 350 = 0.039 in. This means that objects appearing 0.039 in in height, a little more than ⅟₃₂ in, could be distinguished as individual objects, or resolved on the face of the tube in the vertical direction. But can the eye distinguish such a small object when viewing the picture tube from an average viewing distance?

WHAT THE EYE CAN RESOLVE, OR DISTINGUISH

Under ideal conditions of ambient light, contrast, etc., the eye can distinguish line pairs that subtend an angle with the eye of as little as 1-minute of arc, as shown in Fig. 10-4. Under average conditions, the eye has no difficulty in distinguishing objects that subtend an angle of 3 minutes of arc. Suppose (as depicted in Fig. 10-5) we relate the viewing distance V, the viewing angle A, the tube diameter D, and the object size S, to find the lines of resolution useful to the eye. Inasmuch as viewing distance will be expressed in terms of tube diameter so that all data developed will hold true for any size tube, we will begin by finding the relationship between the screen diameter and the screen height.

1. Aspect ratio of the tv image (Fig. 10-6) is 1.33:1, therefore the width of the image is 1.33h.

   Pythagorean theorem gives
   $$D^2 = h^2 + (1.33h)^2$$ from
   which $D = 1.664h$.

2. The lines of useful resolution, that is, the resolution that the eye can use based on average eye performance, are

   $$\text{lines of useful resolution} = \frac{h}{S},$$

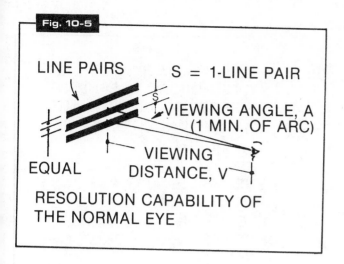

Fig. 10-5

LINE PAIRS   S = 1-LINE PAIR

VIEWING ANGLE, A
(1 MIN. OF ARC)

VIEWING
DISTANCE, V

EQUAL

RESOLUTION CAPABILITY OF
THE NORMAL EYE

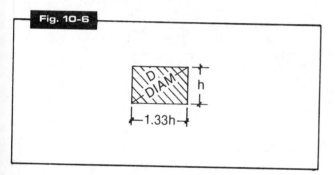

Fig. 10-6

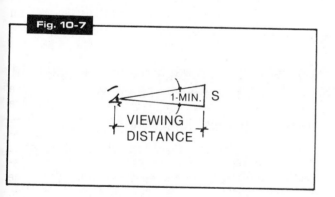

Fig. 10-7

1-MIN. S
VIEWING
DISTANCE

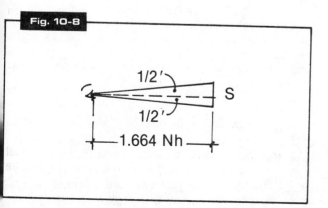

Fig. 10-8

1/2'
S
1/2'
1.664 Nh

where S is found by simple trigonometry according to Fig. 10-7:

$$S = \tan 1' \, (1.664Nh)*$$
$$S = 0.00029088 \, (1.664Nh)$$
$$S = 0.00048038Nh$$

where N is the number of tube diameters in viewing distance.

3. Lines of useful resolution =

$$\frac{h}{S} = \frac{h}{0.00048038Nh} = \frac{2066}{N}$$

4. Repeating steps 2 and 3 for a viewing angle of 3 minutes of arc, we find

$$\text{Lines of useful resolution} = \frac{689}{N}$$

*Note that when we deal with small angles, less than 4°, it is not necessary to draw the bisecting line shown dashed in Fig. 10-8, in order to make a right triangle with a base of ½S. Therefore it is not necessary to write

$$S = 2\tan\tfrac{1}{2}' \, (1.664Nh).$$

We can now compile information for Table 10-1 showing eye performance, when viewing television receivers.

Table 10-1. Eye Performance When Viewing Television Sets

Viewing Distance in Terms of No. Screen Diagonals, N	Viewing Distance From Screen, Ft.						Number of Lines Eye Can Resolve with Angle of Vision	
	Screen Size, Inches, Diagonal						1' Arc	3' Arc
	9"	12"	15"	19"	23"	25"		
2	1.5	2.0	2.5	3.2	3.8	4.2	1,033	345
3	2.3	3.0	3.8	4.8	5.8	6.3	688	230
4	3.0	4.0	5.0	6.3	7.7	8.3	517	172
5	3.8	5.0	6.3	7.9	9.6	10.4	413	138
6	4.5	6.0	7.5	9.5	11.5	12.5	344	115
7	5.3	7.0	8.8	11.1	13.4	14.6	295	98
8	6.0	8.0	10.0	12.7	15.3	16.7	258	86
9	6.8	9.0	11.3	14.3	17.3	18.8	230	77
10	7.5	10.0	12.5	15.8	19.2	20.8	207	69
11	8.3	11.0	13.8	17.4	21.0	22.9	188	—
12	9.0	12.0	15.0	19.0	23.0	25.0	172	—

A 1-Min. of arc is about the smallest angle of vision over which the normal human eye can resolve an object, under ideal conditions of contrast, brightness, ambient light, and signal-to-noise ratio. A 3-Min. of arc "angle of vision" is more typical.

Table 10-1 shows that although the system is capable of producing 350 lines of vertical resolution on the face of the 21-in tv screen, our eye may be able to resolve only 295 with excellent vision, and when seated about 12 ft from the tube. We would be viewing the tube at a distance of 7 tube diameters.

Good viewing takes place at distances between 5 and 10 screen diameters. In large meeting rooms, viewers are often located at 12 or more screen diameters from the receiver. At these longer distances, graphics become difficult to read unless they are specially made with not more than eight lines of captions from top to bottom of the image.

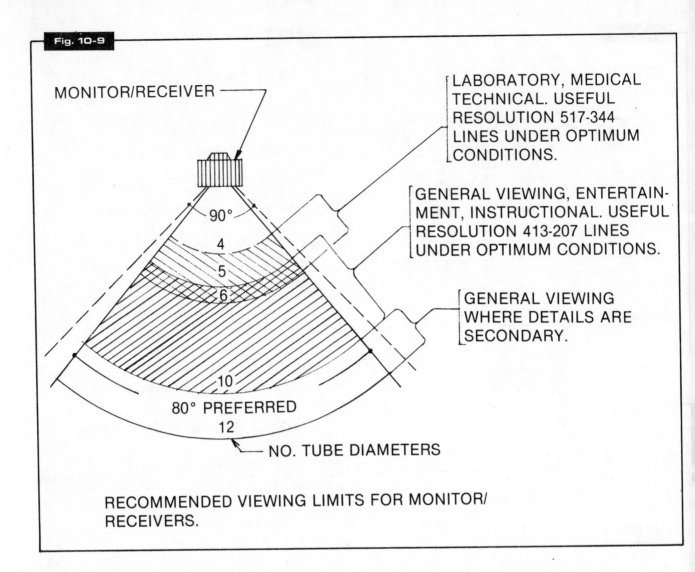

Fig. 10-9

MONITOR/RECEIVER

90°

4

5

6

10

80° PREFERRED

12

NO. TUBE DIAMETERS

LABORATORY, MEDICAL TECHNICAL. USEFUL RESOLUTION 517-344 LINES UNDER OPTIMUM CONDITIONS.

GENERAL VIEWING, ENTERTAIN-MENT, INSTRUCTIONAL. USEFUL RESOLUTION 413-207 LINES UNDER OPTIMUM CONDITIONS.

GENERAL VIEWING WHERE DETAILS ARE SECONDARY.

RECOMMENDED VIEWING LIMITS FOR MONITOR/ RECEIVERS.

From the preceding considerations, it is clear that viewing distance is an important factor when we talk about resolution.

WHAT IS CONSIDERED THE NORMAL VIEWING AREA FOR CLASSROOM OR TRAINING ROOM MONITOR/ RECEIVER VIEWING?

Fig. 10-9 shows recommended areas of coverage for television viewing. The 4-6 tube diameters area serves small groups of up to six viewers who want to see detail and read captions. Often the output of a microscope is televised, using a high-resolution camera and monitor. The 5-10 tube diameters sector, with an 80° viewing field, is the recommended viewing area for general conference room, meeting room, and classroom use.

Table 10-2 gives the actual area in square feet and square meters for the 5-10 tube diameters sector, for both an 80° and 90° viewing field, for the various screen sizes of popular monitors and receivers. Also the number of tablet-arm chairs that would normally fill such an area is listed. Dividing the number of viewers by this figure gives a good idea of how many television sets will be required for proper viewing in a given room. Values are given in English as well as metric units.

WHERE SHOULD TV MONITOR/RECEIVERS BE PLACED IN CONFERENCE ROOMS, CLASSROOMS, TRAINING ROOMS, BOARDROOMS, ETC.?

Figs. 10-10 through 10-16 show some typical situations where television viewing is required. In all cases of monitor/receiver usage, we have shown the 80° viewing angle rather than the often used 90°. This ensures less image distortion for the extreme side viewers. All layouts are within the recommendations of the area coverage in Table 10-1.

The heights at which the monitors are mounted are indicated on each sketch. Note that in the case of the board-room, it was elected not to use television sets, but to make use of a television projector, mounted rear screen to project a large image on the screen. We will talk more about large-screen tv projection shortly.

In cases where television sets are suspended from the slab as in Fig. 10-17, the mount should allow rotation, and the yoke should permit tilting to aim the set at the center of the seating area. A downward tilt does away with bothersome reflection from overhead lights. No viewer should be so close to the set as to have to look up at an angle exceeding 15°.

Fig. 10-10

PORTABLE 54" H. CART OR
FURNITURE CAB. WITH TV & VTR.

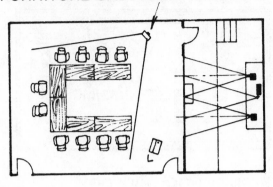

SMALL CONFERENCE ROOM

A. Figure 10-10 showing small conference room.

Fig. 10-11

CEILING OR WALL MOUNTED TV
SET, 7'-0" ABOVE FLOOR, CLEAR.

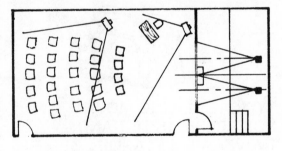

CLASSROOM WITH TWO
MONITOR/RECEIVERS.
MAINTAIN HEAD CLEARANCE.

B. Figure 10-11 showing typical classroom.

Fig. 10-12

TV SETS ON PORTABLE STANDS,
54" HIGH

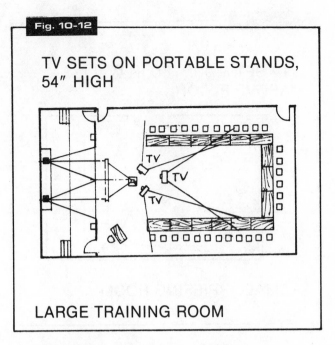

LARGE TRAINING ROOM

C. Figure 10-12 showing training room.

Fig. 10-13

TV PROJECTOR

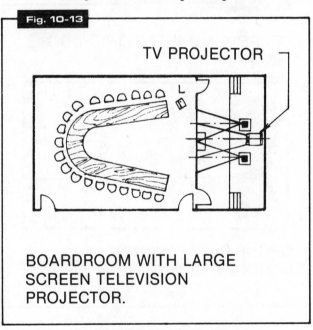

BOARDROOM WITH LARGE
SCREEN TELEVISION
PROJECTOR.

D. Figure 10-13 showing boardroom.

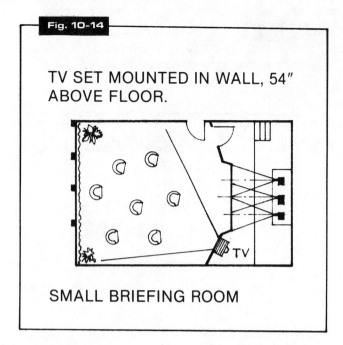

Fig. 10-14

TV SET MOUNTED IN WALL, 54"
ABOVE FLOOR.

SMALL BRIEFING ROOM

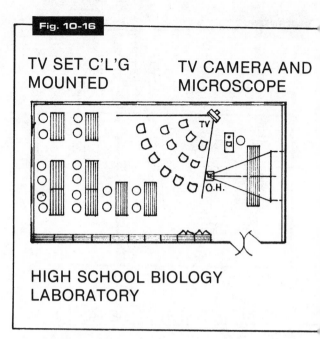

Fig. 10-16

TV SET C'L'G
MOUNTED

TV CAMERA AND
MICROSCOPE

HIGH SCHOOL BIOLOGY
LABORATORY

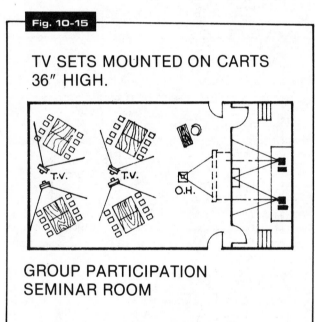

Fig. 10-15

TV SETS MOUNTED ON CARTS
36" HIGH.

GROUP PARTICIPATION
SEMINAR ROOM

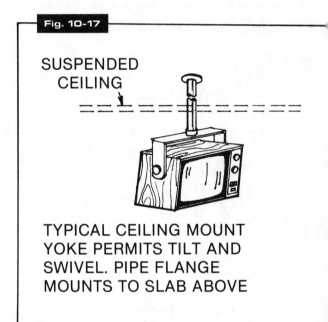

Fig. 10-17

SUSPENDED
CEILING

TYPICAL CEILING MOUNT
YOKE PERMITS TILT AND
SWIVEL. PIPE FLANGE
MOUNTS TO SLAB ABOVE

Fig. 10-18 shows a wall mount, which also must have the capability to tilt and swivel. Wall mounts can be dangerous if not supported properly. Plaster or sheetrock walls on wood or metal studs are not of themselves rigid enough to support the downward torque of a 100-pound television receiver. There have been instances where a set has pulled its mounting right out of the wall.

The possibility of serious injury is present unless steel back-up plates are used, securely fastened to the stud system. Any tv mount should be able to withstand a suddenly applied load of 200 pounds.

When television sets are mounted on mobile carts, the top surface of the cart is tilted forward, or adjustable, to provide 5° of downtilt. Positive means of capturing the set should be provided so it cannot slide forward. The top of such a cart is usually 54 in above the floor, allowing an unobstructed view without head and shoulder interference.

Several manufacturers produce more sophisticated carts that have the appearance of furniture. They usually contain space for one or two video cassette recorders, storage for video tapes, and are available with or without lockable doors.

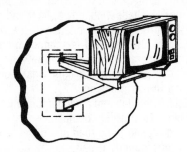

STEEL PLATE BOLTED TO
STUDS, BEHIND WALL BOARD

TYPICAL WALL MOUNT

Fig. 10-18

Table 10-2.
Actual Areas of Coverage for Television Viewing, Based on Viewing Distances of 5-10 Screen Diameters

Screen Diagonal In.	MM	Area of 80° Viewing Sector Sq. Ft.	Sq. M.	Area of 90° Viewing Sector Sq. Ft.	Sq. M.	No. of Tablet-Arm Chairs, Normal Spacing* 80° Incl. View Angle	No. of Tablet-Arm Chairs, Normal Spacing* 90° Incl. View Angle
9	229	29	2.7	33	3.1	4	4
12	305	52	4.8	59	5.5	7	7
15	381	81	7.5	92	8.6	10	12
19	483	131	12.2	147	13.7	16	18
21	533	160	14.9	180	16.7	20	23
23	584	192	17.8	217	20.2	24	27
25	635	226	21.0	254	23.6	28	32
**26	660	246	22.9	276	25.7	31	35

*Based on 8 sq ft per seat, and 0.74 sq meter per seat.
**Some foreign sets have 660-mm dia. tube.

HOW PRACTICAL ARE LARGE-SCREEN TELEVISION PROJECTORS?

Very practical, but until recently their cost has been prohibitive for most AV users. During the past five years, the public has been aware of several brands of "living room" tv projectors, all making use of the dished screen and, generally, a pull-out section of the floor cabinet that supports the screen containing the projection head. Three 5-in diameter projection tubes, red, green, and blue, project a color image, converged to produce a single image on the 50-in screen.

While such well designed units are quite popular, costing less than $3000, their one limitation is that the projected light beam is rather weak, and consequently needs all the help it can get from the gain of the dished screen. However, this light gain occurs on the center and near center viewing axes, but only at the expense of a very noticeable fall-off of light at the wider angles of viewing. These units, although their images are about 50-in wide, are not suitable for large group viewing.

Television projectors have been available for more than 15 years that are capable of extremely large images — 20 or more feet in width — but they require 10,000 watts of power and their cost has soared to a half million dollars! A much smaller unit, and much more affordable, was marketed at a price of 75,000 dollars, but did not compare in image size, brightness, and image quality. Unfortunately there was nothing in between!

In the last three years, a new development in liquid-cooled small diameter television cathode ray tubes permitted the higher voltages needed for tv projection, without overheating the tube. Television projectors using red, blue, and green tubes of this design are now packaged in a unit no larger than a cube measuring 10-in high × 20-in wide × 28-in long, and weighing only 76 pounds. The increased light allows projection onto a flat screen, rear or front projected. Although the manufacturers claim screen widths to 20 ft are possible, the units perform best when the images do not exceed about 8 ft in width. Such a screen size is more than adequate for the average boardroom, conference room, or similar space.

In rear screen use, the projector is suspended from the slab above, over the regular projection table, and is tilted down-

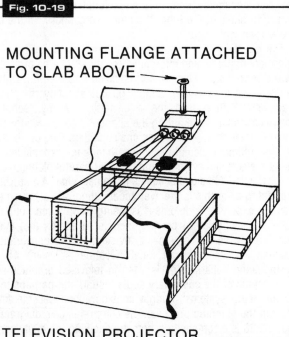

MOUNTING FLANGE ATTACHED
TO SLAB ABOVE

TELEVISION PROJECTOR
MOUNTED FOR REAR SCREEN USE

Fig. 10-19

ward at an angle not exceeding 12°. This does not create vertical keystone, as up to 12° of tilt can be corrected for electronically. The throw distance is equal to 1½ times the image width, so that an image 6 ft wide requires 9 ft of throw distance. This is compatible with standard projection equipment space requirements in AV systems.

The image aspect ratio is 1.33:1, the same as with 16-mm film. At this writing the cost of such a unit is 13,000 dollars.

WHAT KINDS OF INPUT SOURCES CAN BE INTERFACED WITH A TELEVISION PROJECTOR?

The common sources of input to a television projector are live camera, video tape, and off-the-air signals. Note that the camera and tape machine outputs are video, not rf, and are compatible with the projector input. Off-the-air signals are rf, and must be demodulated, that is, changed back to their original form. Demodulation is called *tuning*, hence a tuner is used to select the desired channel before the rf signal can be projected, just as is done within the television receiver.

With the recent three-tube television projectors, the video signal is decoded into its red, green, and blue (RGB) components, each projected separately on the screen, where all three primary color components converge to a single image. For this reason, exact throw distance between lens and screen, and precise aiming adjustment, are important in the initial set-up of the system. Fig. 10-19 shows a typical rear screen mounting arrangement, with the unit suspended from the ceiling slab, above the AV projector table.

Once the tv projector is available in the system, peripheral components may be added to interface such input sources as:

1. Satellite transmission
2. Color graphics
3. Local video conferencing
4. Computer terminal display
5. Remote data line transmission

We cannot escape the fact that this technology is highly system oriented, and it is a mistake to think that all the possibilities mentioned above are readily available or affordable. Careful planning is required when a new facility is in the early stages of design, especially the planning for electrical and related services, and, obviously, space requirements.

WHAT IS INVOLVED IN PROJECTING A DATA TERMINAL DISPLAY ONTO A LARGE SCREEN VIA A TELEVISION PROJECTOR?

The apparent similarity between displaying data on a crt data terminal and displaying the same data on a large screen via projection television is easy to assume. However, the two processes are not similar, except that both displays are viewed on a screen that might be called a television screen.

The data terminal screen (or tube) displays a stream of information that is digital in nature, being the output of a computer. On the other hand the television projector displays a stream of analog information, being the output of a television scanning beam.

Digital information consists of a stream of pulses which, simply stated, is information that has been coded into some

form of binary number system which recognizes two digits, 1 and 0. These digits signify the passage or nonpassage of electrical energy. The pulses are transmitted to a receiver where they are decoded to their original form. Alphanumeric data, linework, symbols, and so forth are readily transmitted in this manner.

Analog information is more akin to photography. Inputs are photographic images — shades, shadows, light and dark objects. These create corresponding voltages at the output of television cameras when the subject is scanned on the faceplate target of a tv camera pick-up tube. The voltages produced are proportional to how much light was seen by the camera lens. Hence the term analog.

Obviously digital information and analog information are not directly compatible. To further complicate matters, the display terminal is designed to be read at a distance of about 15 to 20 inches. Consequently it can display 80 characters per line and 24 or more lines in the vertical direction. Additionally the terminal display can very nearly fill the screen area without losing resolution. By way of contrast, the large television screen is viewed at 2.2 to 8 times its height, requiring a limit of about 14 lines of data, and only 32 to 40 characters per line, excluding spaces between words. Further the television image tends to lose resolution at the extreme left and right sides of the tube, therefore we cannot load the entire screen with data.

The question now is how to transform the digital terminal information into analog information so that it can be projected by the tv projector. This process is called *reformatting*. Unfortunately the user cannot simply go out and buy the "little black box" that does the reformatting. Many technical complications arise requiring the services of a knowledgeable consultant or contractor. When one considers the possibilities of the ultimate communications handling capabilities of today's electronic systems, the need for careful planning and cost-effective decisions becomes extremely vital. A disturbing thought in all of this is the ever accelerating rate at which extremely expensive equipment is being supplanted (but not necessarily obsoleted) by new models. It seems that every month a new version of a major device makes its appearance.

When we talk about an initial investment of some 45- to 60-thousand dollars to get started in televised graphics and data, on top of the customary audio-visual, and perhaps teleconferencing systems, budget considerations become foremost in the planning process. The comprehensive diagram in Fig. 10-20 is a composite schematic showing the many capabilities of a complete communications facility. All of the devices shown would not necessarily be used in any one system, but the system shows how the various kinds of inputs would be interconnected.

When we contemplate video-tape recorders, computers, graphics generators, and so forth, it is natural to think in terms of the actual components. We tend to overlook the fact that there is a need for some kind of system that permits the various components to function together — to be compatible — to ensure final system performance. Equipment system planning must include some consideration for switching devices. The diagram in Fig. 10-20 shows two such devices: a computer input switcher and an audio/video switching device. These switchers act as the traffic signal that routes the selected traffic (streams of communications data) to the various outputs. With a proper switching system, all inputs are fed into the system at the proper levels and in the correct format, before being switched to a desired output.

We have been talking mainly about system "hardware," but of equal importance is the system "software." The software is the information, instructions, and procedures that are programmed on floppy disks, or magnetic tape, which tell the microprocessor exactly what to do to process input data and display the output for any given program. For example, if an operator wishes to input financial data via the keyboard of his terminal, and produce a bar chart on his monitor, with selectable coordinates and colors, there must reside in the system the necessary software program to process the data and display the output in the desired format. The most popular means for storing data and programs is the 8-in diameter floppy disk, used in disk drive units that form a part of the system. These disks are made of a plastic film coated with iron oxide, and are furnished in a protective sheath, which is inserted into the disk drive unit together with the disk. A radial window slot in the sheath allows reading of the magnetic recording on the disk by the pick-up head in the drive unit.

Software is so important for the overall task for which the computer system is to be purchased that it is usually the first item to be considered. All computer system components must be compatible with the software that the user will be able to obtain.

Referring to the schematic diagram of Fig. 10-20 once again, note that a teleconferencing input and a videoconferencing input are included entering the AV switcher. These terms are described next.

WHAT IS TELECONFERENCING?

Teleconferencing, in the original use of the word, denotes a telephone system wherein two-way communication between separated or distant groups is accomplished. It is an extension of the familiar telephone company "speaker phone" system, which featured a small desk-mounted speaker/microphone unit connected to the regular telephone instrument. In operation, a call from one group was placed in the usual manner, and on connection with the called phone the caller hung up his handset and transferred the connection to the speaker phone unit with the push of a button. Several people could now talk and listen, hands off, provided they stayed within a few feet of the speaker phone. The same applied to the receiving end of the line.

The obvious disadvantage of this arrangement was limited group size, and a not very high quality of audio reception. Today's teleconferencing systems recognize the need for conference and boardroom use, with a separate microphone for each participant, and high-quality sound. While the outgoing call is initiated in the usual manner, when the connection is made a switching unit connects the phone circuit to the house audio system in the meeting room, and the conference begins.

The previous arrangement is limited to a single telephone line at location A talking to location B. Several participants can, of course, talk from each location — one voice at a time.

When location A wants to communicate with several locations B, C, D, E, etc., all interconnected, a more sophisticated system is required, making use of what is called a conference bridge.

The conference bridge is more than a piece of telephone gear, it is a manned control center that allows a conference moderator to control the progress of a teleconference. Conferences of several hours duration can thus be held with results comparing favorably with a face-to-face meeting.

The bridge equipment facility can be a service offered by the telephone company, it can be leased, or it can be owned by the host corporation. A complete conference service, including equipment, experience, and personnel, can be contracted to "run" the proposed conference.

Obviously the previous choices will vary greatly in cost, conveniences, and advantages. The cost effectiveness of each scheme requires careful analysis.

Many times, in more sophisticated systems, the meeting room is equipped with a built-in, table-concealed voice enhancement system, and where this exists, the teleconferencing can be made a part of it.

WHAT IS VIDEOCONFERENCING?

Videoconferencing, often referred to as teleconferencing, adds the two-way televised image of individual talkers, or even a group image, to the teleconference. This system is more exotic than straight voice transmission. There is much analysis work and system activity going on today in the field of videoconferencing. Although this form of communication has been in use for 15 years as the "picture phone" system, the techniques of today's television technology have brought about many changes.

One technical problem has always been the need to have the active camera, televising the talker, located as close to the return image of the other participant as possible, so that direct eye contact between the two communicators appeared normal. This means that a camera cannot be placed high, in one corner of the conference room, and successfully televise different speakers seated around a table, while they are talking to the person whose image appears on a centrally located screen. This has the effect of parallax in eye contact, such as occurs when an announcer on camera reads his script from a prompting device located off camera. In the television studio, this noticeable misalignment of the eye contact axis is easy to correct, by the use of a 45° transparent mirror positioned directly in front of the lens, so that both the camera lens and the reader's line of sight are aligned on the same axis. The prompter is mounted below the lens, with its script image facing up. The lens sees the talent through the transparent mirror, via the same optical axis along which the talent reads the reflection of his script.

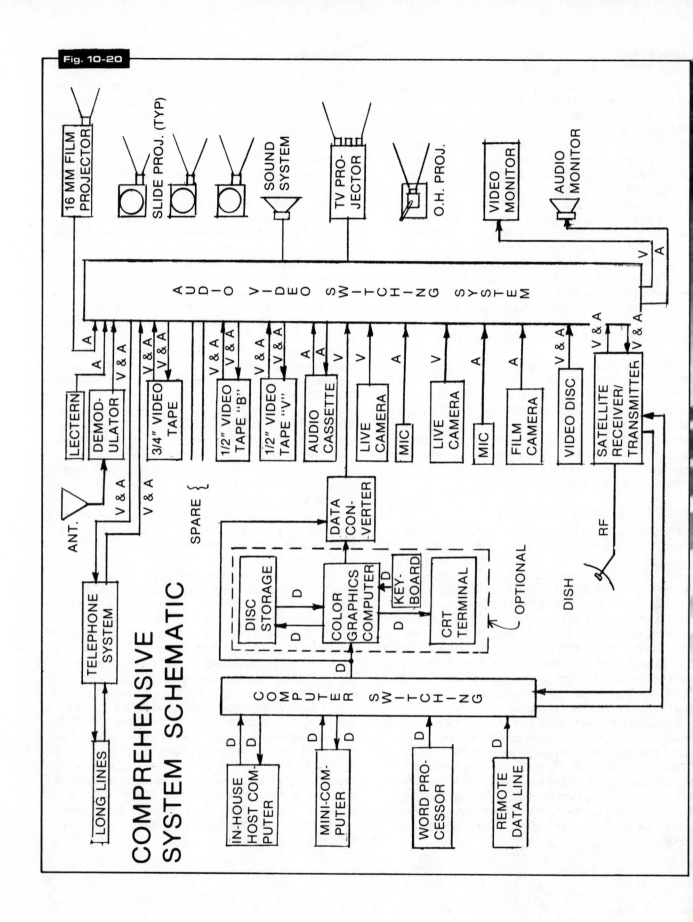

Fig. 10-20

COMPREHENSIVE
SYSTEM SCHEMATIC

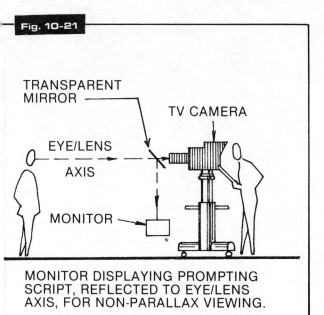

Fig. 10-21

TRANSPARENT MIRROR

TV CAMERA

EYE/LENS AXIS

MONITOR

MONITOR DISPLAYING PROMPTING SCRIPT, REFLECTED TO EYE/LENS AXIS, FOR NON-PARALLAX VIEWING.

While 100% parallax correction cannot be achieved in the normal videoconferencing setup, satisfactory results are obtained by locating the taking cameras as close to the viewing screen as possible. It should be mentioned here that groups of conferees not exceeding eight in number can confer with optimum results. Larger groups create camera pick-up problems.

Aside from some technical and space deployment problems, there is the consideration of transmission costs. For short, line-of-sight distances, microwave or infrared transmission can be used. For intra-building systems, cable transmission is practical. But for long distances, city-to-city or country-to-country transmission satellites are used. Obviously, satellite transmission is expensive; up-links and down-links are required, as well as ground station equipment. The cost of leasing transponder time, or actually owning a satellite, entails very large expenditures. Another consideration is security. Anyone with ''ears'' can listen in, unless a means of encrypting the signal is used.

It is not always the equipment itself that requires large budgets, that is, the hardware we put in our meeting rooms, but it is the system that makes it all work. In any discussion of videoconferencing, it should be remembered that of all the long-distance meetings that take place, only a small percentage of them could profit from making use of videoconferencing. Some meetings demand face-to-face discussions and person-to-person interaction.

In summary, at least for the next few years, it appears that teleconferencing will find more affordable use than will videoconferencing. In the meantime, development will continue, and, hopefully, less costly means of transmission will be found.

Another technique, now starting to emerge, will be the interaction of computer information as a vital part of both teleconferencing and videoconferencing systems.

119

AV CONTRACTING AS A BUILDING TRADE—RESPONSIBILITIES OF THE AV CONTRACTOR

When we examine the spectrum covered by the term audio-visual contractor, we find that it includes many disciplines, as well as many different kinds of contractors. Starting at one end of the spectrum we have the small AV dealer who started a business supplying educational and audio-visual equipment to schools, churches, meeting halls, local business firms, and so forth. At the other end of the spectrum is the large contractor, who, with his resources of men, equipment, and knowledge, has largely been responsible for elevating audio-visual contracting to the status of a bona fide building trade.

Somewhere between these two extremes are found the pure sound-system contractors, the sound-system contractors-turned-audio-visual contractors, the manufacturers-turned contractors, equipment dealers-turned contractors, and brokers who bid on jobs and then subcontract the work. The "complete package" concept has caught on in recent years, and many organizations in the previously named categories have added television, computer, and electronics experts to their staffs, and offer complete consulting, design equipment sales, installation, and service contracts.

Obviously, all the preceding types of practitioners have responsibilities — and these responsibilities may differ widely — depending on where the observer sits. If we tried to examine each situation, we would no doubt end up with something that resembles a code of ethics.

The responsibilities we want to discuss are what contracting is all about. We are talking about audio-visual contracting in the fullest sense of the word. As a building trade, AV contracting opens up new opportunities — and risks — for the contractor, be he large or small. Understanding his responsibilities and the role he must play will take him a long way toward a successful operation.

Here are some comments, remarks, and observations that might be enlightening:

1. There is a substantial volume of building projects on the boards of hundreds of architects across the nation. These projects include corporate headquarters for many of the country's best-known corporations, banks, institutions, hotels, conference centers, educational facilities, airports, auditoriums, and many others. These projects are under consideration abroad, as well as at home. All of them will, no doubt, require the services of an audio-visual/television contractor.

2. Large scale projects like these usually include an independent architect (or group of architects). In some instances large construction corporations are the designer-builder, however, they employ their own in-house architects. In any event, the AV contractor is far removed from the owner, and will probably never see him (or the group which represents him). It is unlikely that the AV contractor will even have an opportunity to meet the architects on a large project. His contacts usually are the general contractor, and the audio-visual consultant.

3. The architect is usually responsible for assembling a team of consultants to assist him. These might include structural, electrical, lighting, audio-visual, plumbing, air conditioning, sheet metal, food, traffic, and landscaping consultants. Architects rarely engage contractors as consultants. That is because the consultants write specifications so that competitive bids can be obtained. If a contractor acted as consultant, he would then be bidding on his own specifications, which is taboo. Competitors would refuse to bid.

4. It is the job of the audio-visual consultant, not the audio-visual contractor, to prepare complete designs, working drawings, specifications, and bid documents, and, in some cases, recommend qualified bidders.

5. Specifications prepared by the AV consultant will often call for equipment *other* than that which the contractor is franchised to sell. Also, methods of designing certain electronic circuit functions, or projection techniques, may differ from those familiar to the contractor. A contractor may handle a glass projection screen manufactured by the ABC company, but the consultant may specify an XY screen. The contractor, if he is interested in submitting a responsive bid, must bid the job *as specified,* but in all fairness, he may submit alternate equipment and methods. In fairness to all bidders, it is obvious that bids cannot be judged unless all bids are based on "as specified" equipment. Only then can the exceptions be taken into account.

6. The contractor must be fully aware of trade union regulations in the site city. For example, failure to recognize that he is not allowed to pull cables or run interconnecting conduits, but must pay scale wages to have the work subcontracted, could be a sizeable setback. This means that before bidding, the contractor should visit the site, and make all necessary arrangements with the general contractor for solving such problems. Similar precautions should also be taken to ensure that certain equipment assembled in the contractor's shop, and shipped to the job, will be accepted on the site. Arrangements must be made to store and protect this equipment until it is turned over to the owner.

7. In most cases a performance bond is required. The ability to obtain a bond should be pursued early in the bidding stage.

8. An item that represents an important responsibility of the contractor is the requirement to furnish the owner complete operating manuals and equipment service manuals, along with complete ''as wired'' and ''as built'' drawings, showing all cable runs and terminations, with all conductors numbered and identified in the rack wiring terminal boards. The contractor should take this requirement seriously, as he cannot receive final payment installment until this requirement is met. Many contractors underestimate the cost of writing manuals and preparing drawings. Some do not have a draftsman or writer on their staff.

9. Another requirement is that the contractor shall instruct the owner's designated personnel in the use of the system. It is suggested that, to protect himself against the cost of repetition by having to give his instruction twice (if the proper personnel were not available), the presentation should be taped. Installed equipment can usually be used for this purpose. Repeat sessions can then be played as often as necessary by the operator.

10. Finally, there is a checkout in the presence of the owner and the consultant. The contractor should be prepared to make all necessary test readings, furnish all test equipment, and demonstrate that all functions are operational as specified.

In summary, it might be said that large job audio-visual contracting is fraught with difficulties, risks, and uncertainties. But so is all large job contracting.

One fear a contractor might have when bidding large jobs is that some competitor, who doesn't comprehend the implications of some of the specified requirements, will, through ignorance, put in a low bid and get the award. This is unlikely, inasmuch as all bidders are bidding the same specifications. Obvious discrepancies and omissions are screened out by the consultant, and only competent bids are considered.

For some newcomers to large job AV contracting, learning to work with consultants, conforming to specifications written by others, coordinating with trade unions and with the work of other trades, taking on financial obligations, and fighting competition may appear too formidable a task. The risks are high, but so also can be the rewards.

Index

IF YOU'RE IN THE VIDEO BUSINESS...

And have a line of video products you need to sell, Sams can help you do it with your own AV/Video catalog.

It's neither difficult nor expensive to have a catalog produced with your name on the outside and your product lines on the inside, and it definitely makes you look good while it helps you sell.

Call 800-428-3696 or 317-298-5566 and ask the Sams Sales Manager for full details.

Many thanks for your interest in this Sams Book about the world of Video. Here are a few more Sams Video Books we think you'll like:

You can usually find these Sams Books at better bookstores and electronic distributors nationwide.

If you can't find what you need, call Sams at 800-428-3696 toll-free or 317-298-5566, and charge it to your MasterCard or Visa account. Prices subject to change without notice. In Canada, contact Lenbrook Industries Ltd., Scarborough, Ontario.

For a free catalog of all Sams Books available, write P.O. Box 7092, Indianapolis, IN 46206.

THE HOME VIDEO HANDBOOK (3rd Edition)
Easily the nation's most popular, most respected book on the subject of home video recording! Shows you how to simply and successfully enjoy your home TV camera, videocassette recorder, videodisc system, large-screen TV projector, home satellite TV receiver, and all their accessories. Tells you how to hook everything up to make it do what you want it to do, how to buy the best equipment for your needs, how to make your equipment pay for itself, and more! By Charles Bensinger. 352 pages, 5½ x 8½, soft. ISBN 0-672-22052-0. © 1982.
Ask for No. 22052 . . . .**$12.95**

THE VIDEO GUIDE (3rd Edition)
If your work involves a hands-on understanding of video production hardware, this is your book. Tells you about standard and state-of-the-art videotape and VCR units for studio or portable use, industrial and broadcast cameras, videodiscs, editing systems, lenses, and accessories, and how they work. Then it shows you how and when to use each one, and how to set up, operate, maintain, trouble-shoot, and repair it. A classic reference guide for video professionals and ideal for those just learning, too. By Charles Bensinger. 264 pages, 8½ x 11, soft. ISBN 0-672-22051-2. © 1983.
Ask for No. 22051 . . . .**$18.95**

THE VIDEO PRODUCTION GUIDE
Helps you professionally handle all or any part of the video production process. Contains user-friendly, real-world coverage of pre-production planning, creativity and organization, people handling, single- and multicamera studio or on-location production, direction techniques, editing, special effects, and distribution of the finished production. Ideal for use by working or aspiring producer/directors, and by schools, broadcasters, CATV, and general industry. By Lon McQuillin, edited by Charles Bensinger. 352 pages, 8½ x 11, soft. ISBN 0-672-22053-9. © 1982.
Ask for No. 22053 . . . .**$28.95**

CABLE TELEVISION (2nd Edition)
Designed for the engineer or technician who wants to improve his knowledge of cable television. Helps you examine each component in a cable system separately and in relation to the system as a whole. Discusses component testing, troubleshooting, noise reduction, and system failure. Contains valuable information concerning fiber optics and communications satellites. By John Cunningham. 392 pages, 5½ x 8½, soft. ISBN 0-672-21755-4. © 1980.
Ask for No. 21755 . . . .**$13.95**

THE SATELLITE TV HANDBOOK
An easily read, amazing book that tells you what satellite TV is all about! Shows you how to legally and privately cut your cable TV costs in half, see TV shows that may be blacked out in your city, pick up live, unedited network TV shows that include the bloopers, start a mini-cable system in your apartment or condo complex, plug into video-supplied college courses, business news, children's networks, and much more! Also covers buying or building and aiming your own satellite antenna, and includes a guide to all programs available on the satellites, channel by channel. By Anthony T. Easton.
Ask for No. 22055

BASICS OF AUDIO AND VISUAL SYSTEMS DESIGN
Valuable, NAVA-sanctioned information for designers and installers of commercial, audience-oriented AV systems, and especially for newcomers to this field of AV. Gives you a full background in fundamental system design concepts and procedures, updated with current technology. Covers image format, screen size and performance, front vs. rear projection, projector output, audio, use of mirrors, and more. By Raymond Wadsworth. 128 pages, 8½ x 11, soft. ISBN 0-672-22038-5. © 1983.
Ask for No. 22038 . . . .**$15.95**